Graham Nerlich is one of the most distinguished contemporary philosophers of space and time. *What spacetime explains* brings together eleven of his essays in a single carefully structured volume, dealing with ontology and methodology in relativity; variable curvature and general relativity; and time and causation. These essays argue that space and time are comprised in spacetime, that spacetime is real, and that its structure forms a main part of the apparatus of the explanation of science. Professor Nerlich provides a general introduction to his collection and also introductions to each part to bring the discussion up-to-date and to draw out the general themes.

T0276002

What spacetime explains

Metaphysical essays on space and time

What spacetime explains

Metaphysical essays on space and time

GRAHAM NERLICH

Professor of Philosophy, University of Adelaide, South Australia

CAMBRIDGE
UNIVERSITY PRESS

CAMBRIDGE UNIVERSITY PRESS
Cambridge, New York, Melbourne, Madrid, Cape Town, Singapore, São Paulo

Cambridge University Press
The Edinburgh Building, Cambridge CB2 8RU, UK

Published in the United States of America by Cambridge University Press, New York

www.cambridge.org
Information on this title: www.cambridge.org/9780521452618

First published 1994
This digitally printed version 2007

A catalogue record for this publication is available from the British Library

Library of Congress Cataloguing in Publication data

Nerlich, Graham, 1929–
What spacetime explains: metaphysical essays on space and time/
Graham Nerlich
    p.   cm.
Includes bibliographical references and index.
ISBN 0 521 45261 9 (hardback)
1. Space and time.   2. Relativity (Physics)   I. Title
BD632.N47 1994
115 – dc20   93–27336   CIP

ISBN 978-0-521-45261-8 hardback
ISBN 978-0-521-04403-5 paperback

For Margaret

Preface

These eleven essays were published between 1979 and 1991 in the following order:

'What can geometry explain?' (1979) *British Journal for the Philosophy of Science* 30: 69–83.

'Is curvature intrinsic to physical space?' (1979) *Philosophy of Science* 46: 439–58.

'How to make things have happened' (1979) *Canadian Journal of Philosophy* 9: 1–22.

'Can time be finite?' (1981) *Pacific Philosophical Quarterly* 62: 227–39.

'Simultaneity and convention in special relativity' (1982) in Robert McLaughlin (ed.), *What? Where? When? Why?* Dordrecht, Reidel: 129–53.

'Special relativity is not based on causality' (1982) *British Journal for the Philosophy of Science* 33: 361–82.

'What ontology can be about: the spacetime example' (1985) *Australasian Journal of Philosophy* 63: 127–42.

'On learning from the mistakes of positivists' (1989) in J. E. Fensted, I. T. Frolov and R. Hipinen (eds.), *Logic Methodology and Philosophy of Science*. Amsterdam, Elsevier: vol. VIII, pp. 459–77.

'Motion and change of distance' (1989) in J. Heil (ed.), *Cause, Mind and Reality*. Dordrecht, Kluwer: 221–34.

'How Euclidean geometry has misled metaphysics' (1991) *Journal of Philosophy* 88: 169–89.

'Holes in the Hole Argument' (1993) in D. Prawitz and D.

Westersthål (eds.), *Logic, Methodology and Philosophy of Science.*
Dordrecht, Kluwer: vol. IX.

I acknowledge the help of the original publishers of these papers whose permission to reprint them here made this volume possible. I acknowledge also the help and stimulus of Andrew Westwell-Roper and his permission to reprint §2, 'What ontology can be about', which we wrote jointly.

I have changed several of the essays in minor ways to update them in some respects.

Birgit Tauss, Karel Curran and Margaret Rawlinson have helped in various ways in preparing the papers for republication. I thank them for their work, patience and good humour.

Introduction

Space, time and spacetime: these are the best candidates for entities which are real, which are central in scientific explanation, but which, in principle, elude any direct observation. Spacetime fuses space and time. It founds the system of spatio-temporal relations which unifies everything of interest to science and common sense; to be spatio-temporal is the touchstone of the real. But space, time and spacetime also seem ideal candidates for bogus entities, plausible but unwanted parasites which the body of ontology is prone to host. To change the metaphor, they look very like hirsute outgrowths on the face of metaphysics which it is the genius of Ockham's Razor to shave smoothly away.

These essays argue that space and time are comprised in spacetime, that spacetime is real, that its structure forms a main part of the apparatus of explanation of science and of plain good sense, but that it may nevertheless confuse us in more ways than one. No serious inquiry into this topic can avoid an involvement with geometry and the spacetime theories of modern physics. Some of these essays look quite searchingly at foundations, particularly of the theory of special relativity, the first and best-confirmed of the theories of Albert Einstein. But surely no one would mistake my work for that of a mathematician or physicist. There is a spattering of sentences in formal notation, but I have been at pains to include but few. Many of the concepts of mathematicians and physicists are of deep philosophical interest, but I have tried to make what I say presuppose as little of this as is consistent with the tolerably brief treatment of the topics which my arguments demand. On the one hand, I like to remind myself of where I think the more technical ideas spring from and, on the other, I want the company of every philosophical reader who might wish to follow me.

All my arguments aim to support, in their different ways, a general thesis about spacetime which is best called *realism* (though sometimes called substantivalism or absolutism): spacetime is a concrete particular and the practice of scientific explanation and invention firmly commits us to its existence. It is fairly generally conceded, I think, that this is the natural attitude to take to the question whether spacetime is part of our ontology, and I think the concession is a judicious one. On the face of things, and in particular on the face of physics, things, events and processes do commune with one another within the arena of space and time. Perhaps it is not so natural to view spacetime as a particular, concrete entity related in close, concrete and even causal or, better, quasi-causal ways to the material hardware of the universe. Perhaps it is, even, *un*natural because we think of spacetime as eluding perception as a matter of its ontic type and because, classically at least, space was not conceived as a substance, 'because it is not among the proper dispositions that denote substances, namely, actions, such as thoughts in the mind and motions in bodies' (Newton 1962, p.132). But, overall, realism about space is our first response. What this adds up to, if it is correct, is that an initial burden of proof lies not with realism but with the views that oppose it.

In any event the burden of proving realism an error of metaphysics has been shouldered with enthusiasm by most of the philosophers who have considered the issue over the last hundred years. Most of this introduction is devoted to a synoptic picture of that work rather than to a summary of what I have tried to argue against it.

One thinks of a modern period of vigorous debate as beginning with Poincaré's arguments (1952) that we can always choose a Euclidean geometry if we choose to make enough sacrifices of convenience to retain it. However, this will commit us to no sacrifice of the factual content of our story of the world. This begins the view called conventionalism, that no choice of geometry is a choice of spatial facts. There are no spatial facts but only conventions which make our world-pictures more elegant without any possible risk of trespassing beyond the bounds of truth.

By far the most important event in this century as far as this study is concerned happened in 1905 with Einstein's seminal paper (1923, §III) introducing the theory of Special Relativity. The theory postulates that Maxwell's equations for the electromagnetic field take the same form for any inertial frame of reference. This means that there can be no absolute simultaneity: what had always been taken as an unshakeable, *a priori*, synthetic necessary truth is false. It is not the case that for any pair of events they are either absolutely simultaneous or one occurs absolutely earlier than the other. Einstein gave an epistemological argument for the falsity of absolute simultaneity, that any attempt to measure the simultaneity of two events in a frame of reference winds up somehow presupposing what it purports to measure. He went on to claim that simultaneity could be decided only as a matter of definition or convention. There were no matters of fact about it. Although it makes sense to adopt the same *form* of definition in every frame, the result will be that each distinct frame will have distinct classes of simultaneous events. Simultaneity as a relation of objective fact was ushered from the scene.

This event, just by itself, darkened philosophical counsel on the content of special relativity for decades and it probably still does. But this was not all. Minkowski's discovery, in 1908, changed our perspective in another way: special theory could best be interpreted not as the unfolding of physical processes in space and time but as their pattern in spacetime. Space and time were to lose substance and become mere shadows of their classical selves, while a fusion of the two took centre stage as a main actor in the world of physics. In particular, space, the bugbear of metaphysics ever since Newton's powerful and insightful arguments for it in *Principia*, was banished from a place in the ontic sunshine.

Lastly, the theory of general relativity was thought at first to hammer the last nails into the coffin of space or spacetime as a real things. There were two main reasons to think so. First, it seemed that the new theory yielded a principle that motion was generally relativistic: that any smooth time-like path (family of paths) fixed an acceptable frame of reference for physics so that motion, defined just on material things, was a strictly symmetrical relation. Closely connected with this but yielding a somewhat different conclusion, was the fact that the laws of general relativity were generally covariant: the form of the laws was invariant under any diffeomorphic (smooth continuous) transformation of coordinates. Confuse this with the idea that the observables of general relativity can consist only of what does not vary when we distort spacetime by any diffeomorphism and you have the momentous conclusion that the world is definite only up to those properties unaffected by such distortions of spacetime. Only space-time coincidences are real, hard, observable facts. The remnant is only the *seeming* content of physics, a conventional overlay of useful but not factual language.

There were two main consequences of all this for metaphysics. It appeared that the separate streams of philosophy and physics had at last joined in a broad river of discovery and progress. In particular, radical empiricism had enlightened physics as to its epistemology. This liberated it from the chains of *a priori* dogmatism. In turn, physics, in a series of brilliant, creative discoveries, had resoundingly vindicated a variety of empiricist practices and credos which, thus fortified and enlightened, could repay the favour by a yet more stringent and illuminating critique of science. Thus began the vigorous wave of twentieth-century positivism.

A second consequence was more specifically doctrinal. There was broad agreement that progress in physics led (and was led by) a programme of ontic economy, by Ockhamism. There was a series of steps from Newton's absolute space to the classical relative spaces of inertial frames, thence to Special Relativity's relativising of a vast range of physical properties and quantities to these frames together with the dethronement of time from its ancient pedestal, thence to space-time's collapsing of space and time into a single entity, and finally the giant step to general relativity and general covariance. Each step seemed to leave us with a barer, leaner world, the more frugal ontology of which led to increasing sensitivity of theory to observation and of explanation to the structure of the material world. The progress of modern physics was the same as the diminution of its content and especially of its ontic load. This view of twentieth-century physics gained uncertain confirmation from the rather instrumentalist approach to quantum mechanics found in the Copenhagen interpretation.

The decades have made it increasingly clear that nearly all of this was illusory. Although it was not always my intention to address this illusion when I first wrote the essays in this book, it now seems to me that perhaps this has been my dominant (by no means my only) theme. Newton (Cajori, 1947) had a lucid account of why apparent uniform motion is indistinguishable from real uniform motion: space is Euclidean and its symmetries hide the distinction (as §7 shows). The post-Newtonian classical physicists left the ontic status of inertial frames and the symmetry among them obscure. Without the realism of a physical Euclidean space, the ontic foundation of the relativity of uniform motion was anything but transparent. Nothing sustained it. Yet how could it rest on the *absence* of an entity unless it were an entirely general relativity (as so many philosophers were convinced *a priori* that it must be)? From the standpoint of metaphysics this was not a forward step. Special relativity before Minkowski deepened this obscurity since, in an ontology of enduring three-dimensional objects and time-extended processes there is this anomaly: three-dimensional objects have none of their three-dimensional properties (shape, mass or duration – though charge is an exception) well defined intrinsically, despite the fact that each takes its existence, so to speak, as a shaped or massive or enduring thing. The fundamental material entities were continuants, but had no intrinsic continuant (metrical) properties. Each property had to be related to one of an infinite set of privileged frames of reference, yet the basis of this privilege remained mysterious. This was puzzling indeed. Spacetime clears that fog by fusing space with spacetime and swallowing continuants up into four-dimensional objects. But Minkowski spacetime is not less structured than Newtonian or post-Newtonian classical spacetimes. It is richer: where they are defined only up to affine structure, relativistic spacetimes are metrical. Many of the more striking consequences which apparently flowed from generally covariant formulations of spacetime theories were simply the output of confusions. As one would expect, for much the most part the growth of physics reveals that the world is richer than we used to think it.

Confusion over these last points is the background against which the philosophy of conventionalism flourished. It was extended to embrace almost all of the structures which we ascribe to space, time and spacetime. They were seen as outrunning the actual structure of facts in the world so that the metric, affinity and projective structures of spacetime were regarded as fictitious impositions on a basic, smooth topological manifold. Even that minimal structure has been challenged. Reichenbach, the most distinguished of the positivistic conventionalists, argued that the topology of space is a convention (1958, §12). That strand of argument is not considered here (but see

Nerlich 1994, §§7 and 8). It has also been considered from a quite different point of view much more recently by Earman and Norton (1987) and is discussed in chapter 9.

It is at this point that conventionalism joins hands with an older anti-realist strand in the metaphysics of space and time: relationism. This tradition can be traced back to Leibniz, though it is not at all clear that Leibniz would have welcomed the modern claim to his paternity of this group of doctrines. There is no doubt that Leibniz often reads like a modern relationist, but his thought on space and time springs from the philosophy of monads and is, in that respect, deeply opposed to a fundamental tenet of modern relationism. The tenet is this: we can state all the facts and gain every other advantage for science and metaphysics by eschewing all serious talk of space in favour of a theory which confines itself in one or another way simply to material objects and spatial relations among them. These are real and objective, the rest a colourful fiction which should be understood as such. Leibniz insists on the status of space as a mere representation, quite clearly. What brings in doubt the legitimacy of his parenthood of relationism is just that he regards spatial relations, too, as merely well-founded phenomena, as apparent rather than real. But Leibniz certainly argued that, phenomenal or not, the whole system of spatial relations of things to things may be detached from the system of spatial relations of things to space. Thus the latter may drop out of our metaphysical picture of things as an entity which has no role in our intellectual economy.

Certainly some argument to the effect that spatial relations are not themselves tainted by a commitment to a sustaining space seems to be needed. Else it remains uncertain, at best, whether or not the question is begged. On the face of things, even simple spatial relations such as *x is at a distance from y* cannot hold unless there is a path between *x* and *y*; that is, there must be a point *z*, say, between *x* and *y* and a further point between *z* and *x* and so on without end. The plausibility that spatial relations are mediated by space is just the plausibility that if one thing is at a distance from another then there is somewhere some way between them, whether or not the place is occupied by something. That is not a decisive argument but, I submit, it is a highly intuitive and plausible one (see Nerlich 1994, §1, for an extended discussion). As far as I know modern relationists rely on Leibniz's detachment argument to extricate spatial relations from dependence on such spatial entities as paths. That argument is considered in §6.

Of course, there is the obvious theme of the relativity of motion and the consequences for the reality of space or spacetime of its failure – if it fails. That is the topic of §5 of Part 1.

In 1983 relationism was given, independently, two splendidly lucid and carefully worked out formulations which undertook to show how the programme of the general relationist theory might be carried out in full. The two versions, one by Brent Mundy and the other by Michael Friedman, differ in detail but are fundamentally alike. I shall briefly describe the former here. The latter is described in §9.1.

Consider just the set of all physically occupied spacetime points. Let the spatial relations among them be comprised in quantities which provide an inner product structure among the members of the set. Mundy shows how to construct from this information a partial but concrete 4-vector-space structure for the set of occupied points. This, in turn, enables him to show how to embed the material structure in the complete abstract vector space of real numbers. He shows that this embedding is unique up to an isometry. The point of the unique embedding is to justify the use of the well-understood abstract vector space to describe and predict results for the embedded concrete related points. The full vector space functions simply as a representation. Mundy concludes that it is a mistake of metaphysics to regard it as itself a part of the real concrete world.

This view is considered at some length in §1 of Nerlich (1994). A few observations on it are relevant here. It has the unquestionable advantage that its basis is identical with our epistemological basis for such knowledge as we have of spacetime. Its weakness is, perhaps, that whereas older relationist theories tried to show that space, time and spacetime were objectionable in themselves, involving some absurdity or metaphysical confusion, this version of relationism is committed to the view that spacetime, complete with unoccupied points, is a perfectly intelligible structure. That is how it can be of use to embed the material structure within it. Otherwise the embedding would be absurd. So there is only one ground on which the relationist's reduction can be urged: ontological economy. If this ground is to support the argument built on it, then the role of the full abstract vector space must surely be confined to that of an instrument of calculation. The theory which the embedding allows us to exploit functions purely as a measure space. Although Mundy does not seem to say that quite explicitly, I believe that it is part of his thesis that the role of the representing space is confined to that.

Much that I have to say argues for the view that the role of space-time in physical thinking is very much richer than that. Parts 1 and 2 are largely devoted to spelling out in what way spacetime is a richly structured, explanatory concept. No less importantly, I try to show how considering it as a structure independent of its occupancy has allowed some physicists to invent interesting, speculative theories about deviant ways in which it might be occupied. Deviant, that is, from the convictions about occupancy which originally shaped space-time theory.

The more recent relationist approaches, ingenious, careful and rigorous though they are, repudiate much of interest in the older traditions of relationism. First there is the argument which, in its very crudest form, holds that space is unreal because it is imperceptible as a matter of its ontic type. Often, surely, and without explicit mention, it has served to motivate the search for anti-realist arguments. In a more sophisticated form, it founds the variety of arguments for the conclusion that space, time and spacetime have no explanatory role to play because they are in principle unobservable. I scrutinise some of these older arguments.

There is a prevailing view that it is causal relations which are observable and which really play the role that realists think is played by spacetime relations. That view, I think, genuinely does derive from Leibniz. The clearest and most interesting grounds on which to argue this spring from one or another form of the causal theory of time: that temporal relations are really causal ones. I oppose this causal theory. It is in this way that the essays which deal with time by itself enter our field of view. This is in turn tied to the very deep-seated, highly intuitive yet, in the end, very puzzling metaphor of the flow of time. That the metaphor is a profound (if imperfectly lucid) way of characterising time is the subject of a vast, intricate and extraordinarily difficult area of philosophy. I argue that the metaphor is a deep confusion. Time does not flow, nor is there a becoming. The occurrence of events is not importantly different from existence for objects, but a counterpart of it. Not only is the analysis of what founds the impression (whether illusion or perception) that time flows both intricate and hard, the sense of conceptual stress among writers in the field is marked. Everyone's view of time is, at some point in it, deeply counter-intuitive. This goes a long way toward explaining the strenuous excitement with which one reads (and writes) essays on the topic. I share the view, conceptually strenuous enough, to be sure, that time is essentially like space in its ontology: there is no ontologically privileged time, the present, any more than there is an ontologically privileged present place.

My essays on time are addressed to the sources of conceptual stress rather than to the profoundly difficult questions about what is the basis of the fact that at least temporal phenomena are somehow asymmetrical. I touch on these issues but you should not look here for any additional insight into that basis.

These are the main threads that tie these essays more or less tightly together. There is an interest or an emphasis which might make them peculiar in more senses than one: my concerns are more centrally those of a metaphysician than of a philosopher of science. Perhaps this lends almost an air of naivety to some of the essays, especially in Part 3, which neglect questions which are of concern to most philosophers of science. My main aim in them is to address general areas of conceptual stress and to avoid loading readers with details which, however fascinating and difficult in themselves, are not crucial to the sources of stress. I should mention one topic which is prominent in the philosophy of science literature on spacetime which figures only very incidentally in this work. It is the question whether, and to what extent, spacetime theories are underdetermined by the observations on which they depend. My interests are those of the ontologist rather than those of the philosopher of science in general. Where epistemology intrudes, it is seldom brought to bear on the question of underdetermination. Apart from the focus on ontological issues there is an important reason for this neglect which I shall mention just briefly.

The idea that spacetime theories are underdetermined springs from a quite particular view of what the observations are. The observational basis for spacetime theories is supposed to lie just in the set of material spacetime coincidences (Friedman 1983, pp. 22–31; Sklar 1974, pp. 115–17). The view is implausible in itself and motivated essentially by confusion. It is implausible since we judge rather accurately, and have done since we were young and naive, such things as that coins are round and fit in pockets, that they don't change their shape and size noticeably under ordinary conditions, that most books are not round and that they are bigger than coins, few fit in pockets and most are smaller than the people who read them. These humdrum judgements are too strong for the interesting versions of conventionalism to swallow as matters of fact. Few of them judge mere spacetime coincidences. All of them are observational. The view is motivated by the confusion mentioned before about the fact that our theories can be formulated covariantly. That the laws of spacetime theories are unchanged in form by diffeomorphic transformations of coordinates does not mean that spacetime and physical structures are real only in as far as they are invariants of active diffeomorphic distortions of spacetime into itself. It is not even possible for us to imagine, either by visualisation or by kinaesthetic fancy, what it would be like to encounter a thing determinate only up to a smooth topological structure – a closed curve, say, that was neither circular nor variably curved, altogether without adornment from projective, affine or metric structure. So far is it from being the case that diffeomorphic invariants exhaust the content of observation that we cannot even visualise a world with so impoverished a structure. Of course, that appeal to the limits of our imaginative prowess is no proof that the stuff of observation must be metrical. It does not mean that we can't *understand* what it would be for things to be metrically indeterminate. But it surely places an onus of argument squarely on those who would so sharply, so implausibly, impoverish observational content. The underdetermination of spacetime metric by observation is a myth.

Part One

Ontology and methodology in relativity

The essays in this part are about issues which arise within special relativity in one way or another. Each of them argues that the special theory is a more structured, ontologically richer and deeper theory than many philosophers have taken it to be. Several of the essays raise questions of methodology for the metaphysics of spacetime. All of them sustain and explore the theory that spacetime performs a unique and crucial role in explanation which earns it a secure place in our ontology.

From rather early in its development it was realized by Robb (1914) that the full metric structure of spacetime could be built on surprisingly slender foundations. Robb laid down a few axioms on a relation which he called *after*. This idea failed to seize general attention for a number of years. It was taken up subsequently by a number of writers, notably, for our purposes, Zeeman (1964), Hawking and Ellis(1973) and Winnie (1977). (A compact, accessible and elegant account of this may be found in Lucas and Hodgson (1990), chapter 3.) The key relation was now called causality by all these writers and is now thoroughly established in the literature under that name. This lends colour to the idea that, somehow or other, a notable reductive victory has been achieved: the geometry of spacetime has been dethroned from its status as an abstract structure within which material structures are contained. The only structure we need to be concerned with is fully physical and unpuzzling. It is causality.

A recurring theme among the essays of this part is that nothing of the kind has occurred. I must emphasise that I don't dispute the main substance of the work of Robb, Zeeman and the others. Far from it. It is a central part of my case that their formal work is sound, elegant and deeply insightful into the basic structure on which Special Relativity is built. I quarrel only with the opinion that the relation on which the formal work is based is causality. I try to show that it is, in fact, a geometrical relation, a structure in spacetime and not founded or dependent on material relations at all. It is the relation which divides elsewhere from elsewhen. Robb's intuition was sound in calling it 'after'.

A main thrust of this argument is that the Limit Principle – that nothing outstrips light or that light is the limiting speed for propagation – is not a core principle in SR. It is more precise to choose the Invariance Principle: that the laws of electromagnetism are form-invariant under transformations of the Lorentz group. But, I must stress that not even that is quite accurate: one needs to define the relevant structure in essentially geometrical terms. It rests on projective and conformal features which can be captured in just the formally elegant ways which the literature so luminously presents. More modestly (in the ecumenical spirit mentioned a little later) one might say that the Limit Principle is not the *only* core principle of SR.

Closely connected with the topic whether causality is the basic relation for SR is the vexed issue of simultaneity for inertial frames of reference. Einstein argued in his 1905 paper which began the theory, that one has to define this relation relative to a chosen frame by means of an arbitrarily stipulated definition. That the speed of light is invariant in every frame is simply the result of that stipulation and not any matter of fact relative to it. This idea dominated the philosophy of relativity theory until late in the 1970s. That simultaneity was no matter of fact animated a dominant view of the philosophical significance of relativity theory. Success in physics could be seen to go hand in hand with various positivistic techniques for reducing its apparent content.

I argue that this was based on several confusions, most notably the confusion just mentioned, that SR presented us with a world shaved by Ockham's Razor, a world in which space and time as things independent of material processes have been shown to be delusive.

Malament showed (1977b) how to settle simultaneity relative to a frame of reference on a basis which was suitably objective, unique and convention-free. As an operational technique it depends on the passing of light signals. I explore other, related, ways of fixing it objectively, uniquely and free of conventions. It is not crucial to the objectivity of the fixing of simultaneity in a frame that the Limit Principle be true either in the sense that light fills the null cones of spacetime nor that there are no faster-than-light particles (as I argue particularly in §1). Among what one thinks of as principles of SR, it is the Invariance Principle which fixes simultaneity unambiguously. But once again, the true basis lies not in the Principle, but in the geometric structure of spacetime: on its conformal and projective features. These, in the case of SR's Minkowski spacetime, define those strong global symmetries all of which can be characterised by the single relation and the rather simple axioms on it which were first noticed by Robb.

There is another way in which the rich structure of Minkowski spacetime can be characterised and which has been misunderstood by positivists because of their convictions and ambitions. It is the question of the relativity of motion. In metaphysics the question has always concerned whether we can define motion as a fully symmetrical relation when we define it on observable material objects. One dominant tradition has been that the question ought to be given an affirmative answer on some sort of *a priori* ground. The basic thought here is, no doubt, that the failure of relativity of motion to be perfectly general makes it difficult to exclude space (or spacetime) from our ontology, yet its admission is abhorrent on *a priori* grounds. The awfulness of geometric entities presumably lies in their apparently principled resistance to any kind of observation: no decent entity can be imperceptible in principle. But rest and motion are, at best, relative only to some privileged class of frames. Empirical study of the observable world has not been wholly kind to radical empiricist doctrine.

I pursue this theme in rather traditional style. I consider what limitations there are on the relativity of motion conceived of as finding a frame of reference which gives us a global picture of a spatial universe evolving over time, considered equally globally. This is somewhat alien to the rather localist approaches to the topic which General Relativity makes natural. But I think that the issue framed as I discuss it, remains important for metaphysics. I use it, in the context of SR, to show how, once again, the richer structure of SR and Minkowski spacetime are illustrated in rather stringent limitations on relativity.

These themes pervade the essays and I return to them in various forms. But there is also a group of methodological preoccupations which emerge more or less prominently and recurrently in them.

The bare question whether spacetime is real can't be extricated from questions of its structure. Structures are certainly ontic. But they don't really consist in *entities* residing in spacetime (though the literature sometimes finds this way of phrasing the matter convenient). The question what structures are real is not readily settled by considering which objects are values of the variables of quantification, as many ontic questions are. So there is an issue of just how this aspect of ontology is to be pursued.

In an ecumenical spirit, I also conjecture that metaphysicians should recognise various aims which philosophical studies might have. This might permit a recognition of a broad common purpose among apparently conflicting viewpoints. We might agree that if we want to find the most economical form of a theory, its Ockhamist core, so to speak, we might arrive at a rather different picture of it from one we would reach if we were concerned with other desirable features of theory. All these pictures add substantially to our understanding of spacetime (thus I sympathise with the remarks in Lucas and Hodgson (1990), pp. 110–12 on some of my less compromising remarks). While I strongly incline toward the view that positivist philosophy of science is largely motivated by a deep misunderstanding of the physics of the relativity theories, it would be quite absurd to deny that nothing of worth ever came of characteristically positivist probes into spacetime theories.

One result of that tolerant approach to things is more crucially important, I now think, than it seemed when I wrote the papers. What I call the permissive core of a theory is that part of it which remains unchanged under the pressures of speculation and invention within the theory. It is what provides the framework for new suggestions about the behaviour of matter. For instance, that matter might not always lie inside the null cones and light not always on its surface. These may well contradict the first ideas about physical processes which led to the invention, formulation and adoption of the theory. But the new ideas are sustained and made systematic and rigorous in what is recognisably still the original theory. As a matter of historical fact, just this has occurred in the history of SR. The development and speculative predictions both about the existence of tachyons (faster than light particles) and the massive photon (which would not, if massive, be propagated in the surface of the null cones) are of just this sort. The framework of SR which invigorates and disciplines this work quite clearly lies in the geometry of spacetime: in the null cone structure which characterises the conformal and projective makeup of Minkowski spacetime. It is quite independent of the relations which happen to hold among the occupied points of spacetime. Thus realism about the geometric structure itself may come to play a central role in a picture of how a scientific theory changes and develops. It constitutes a strong objection to relationism, therefore, and a powerful case for realism.

Finally there is the methodological theme of the unsatisfactory world. I mean by this a world which is certainly logically possible and conceivable but which, because of its structure, does not lend its theoretical ideas elegantly to observational disciplines. One clear case of an unsatisfactory world is Newton's. In one way, it is admirably clear. It tells us precisely how cause and force distinguish uniform from non-uniform motions. It tells us exactly why rest cannot be distinguished from uniform motion. The geometry of space is Euclidean. But it leaves us quite without means to distinguish them in practice. This gave rise to insoluble metaphysical disputes about whether the distinction could make sense if unobservable. Since practice led inevitably to the acceptance of an infinite class of acceptable frames of reference, it was argued that this was a somehow better account of the world metaphysically than the one Newton had given. Somehow it was thought to have got rid of something somewhere in the ontology of physical theory. But without appeal to a spatial entity, the basis of the relativity of motion among these preferred frames remained obscure. Perhaps the quantum world with non-local connections is also an unsatisfactory world.

This idea surfaces in the papers of this Part, but is tied more intimately, I now think, to the papers of Part 2. The aim of some of those papers, at least, is to extend Newton's unsatisfactory world to a closely related but much more satisfactory one, and to use the comparison to show that Newton's world is rather more satisfactory metaphysically than it is generally reckoned to be.

Michael Friedman has urged that we will not learn from the mistakes of the positivists unless we learn to look with a more sympathetic eye than philosophers are wont to cast on the motivation for their work. I take it that Friedman is himself rather sympathetic to the work of Reichenbach because he finds the motivation for it deep and, I diffidently think, because he believes that there was something significantly right about their motivation. Even more hesitantly, I think he believes they saw correctly that the progress of twentieth-century science shows that as we gradually detach our thought from what Kant would have called the formal elements in our knowledge of the world, and attach it more closely to the observable elements, our knowledge becomes leaner, tauter, harder, clearer. We rid ourselves, as we progress, of excess baggage. The march of science coincides with the drive toward ontic economy. My hesitation is based on the fact that his own brilliant and penetrating work in the area does not suggest, itself, that anything of the kind is true.

However that may be, the first essay selects some examples which suggest that a different lesson may be learned from the mistakes of the positivists. The lesson is that, in many cases, their mistakes led us in quite the opposite direction from the truth. The common feature in all of these examples is that the positivists failed to realise that Special Relativity is a richly structured theory. This turned them to tackle the wrong tasks and to tackle them with the wrong methods. That is what we have to learn from the mistakes of the positivists

1 On Learning from the Mistakes of Positivists

1 Aims of the paper

Philosophers have much extended our understanding of the foundations of spacetime physics in this century. They go on doing so. Much of this is the work of positivists and conventionalists. Friedman (1983, Introduction) tells us in his excellent book that we must learn from their mistakes if we hope to go beyond them. I will look at two of their mistakes: one is about simultaneity and the other is about the relativity of motion.

In my view, the most interesting and suggestive arguments take the old-fashioned form that the world could not possibly be as our best scientific theories say it is. One argues that physics, as the scientists hand it down to us, is conceptually awry and must be rewritten, either by amputating parts of it or by interpreting these parts as conventions, not factual claims. They are metaphysical arguments. In this century, Mach, Einstein, Reichenbach and Grünbaum have urged them. Epistemology appears in them only by ruling out concepts from a fact-stating role unless they meet some observational criterion or other.

It would be absurd to claim that this exhausts the normative role of epistemology in the philosophy of science. But I see little value in asking, about these dismembered forms of physical theory, whether they are better evidenced than the standard forms used by the practising physicist. We philosophers are not likely to choose better than they do, even if we enlarge their view of what they may choose among. Further, a main theme will be that a focus on the observationality of ideas and on parsimony of theoretical structure misled us for decades as to what spacetime theories were about. Such preoccupations are still with us. I urge that we need to look, too, at which structures in a theory may help us grasp how it may evolve under the pressures of new evidence and speculation.

I make some large assumptions. No distinction between observation and theory statements will sustain the fashionable thesis that theory is underdetermined by all possible observation statements. I have argued elsewhere (1982) that Quine's pragmatist holism gives no workable account of the deep entrenchment of propositions in theory – the pragmatist surrogate of necessary truth – or of theory change. Also, that Quine confuses the range of observational vocabulary with the range of observable fact (Nerlich 1976). I agree with others who have argued that no strong principle of charity is defensible. I believe that single sentences, as atoms of truth, have been neglected and ill understood by modern pragmatism. I have put these assumptions crudely and briefly, but the aim of my essay will be misunderstood without some brief reference to them.

2 Conventionalism and simultaneity

I turn first, and briefly, to the question of the allegedly conventional determination of simultaneity in SR.

Bridgman's (1962) proposal ought to have closed the debate on the status of simultaneity. Use of slowly transported clocks properly defines a relation which is unique, relative and not arbitrary. This is clear in a spacetime picture; a glance shows that the definition can have no rival. Yet no consensus accepted it. Malament's analogous but more effective presentation of Robb's proof that we can define simultaneity in terms of causality has no advantage of greater transparency, nor of greater fidelity to real causal relations. It succeeded because it did make one, but only one, concession to conventionalism; it was explicitly causal. Thus the core of the objection to conventionalism was perfectly clear.

Let me illustrate how limpid Bridgman's definition is. Choose a reference frame and two rest clocks, *A* and *B*, at different places in it; represent them by two parallel timelike lines in spacetime (fig. 1.1). Choose a point event, *e*, on one of them – *A*, for example, Imagine the trajectories of uniformly moving clocks which might lie as close as you please to *e* and equally close to some point event on *B*. There are two limits to the trajectories of such clocks. First, there is the light ray which intersects *e* and some point event on *B*; second, there are trajectories which approach arbitrarily close to parallelism with the worldlines of *A* and *B* while sharing a point with each. The first limit can offer no satisfactory definition of simultaneity, since different simultaneity relations emerge depending on whether we place *e* on *A* or on *B* (fig. 1.2). We have no way of preferring *A* over *B* for the location of the reference event *e*. But if we choose parallelism with the frame clocks as our limit, then we get the same simultaneity relation whether we place *e* on *A* or on *B*.

Thus we have the ideal situation of a limiting trajectory which approaches as close as you please to *A*'s, as close as you please to *B*'s, and which has a point in common with each. So long as we stick to the assumption that our clocks do, indeed, measure the length of their own worldlines, the definition is unique, non-arbitrary and relative, varying quite obviously with the direction of the worldlines (i.e. the frame of reference chosen). Once we accept the definition, then the

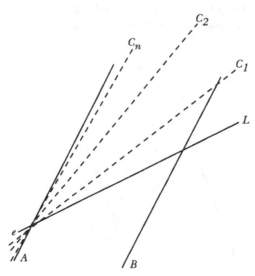

Fig. 1.1.

trajectories in question are clearly those of clocks whose motion is arbitrarily slow relative to the chosen frame. Equally clearly, the resulting relation is satisfactory in being neither absolute nor arbitrary. It is the same as Einstein's alleged convention.

If this is so clear, why was it not seen at once that Bridgman's suggestion settles the issue? Similarly, why was the work of Robb so long neglected? Partly, I suggest, because an argument about the issue was unlikely to be recognised unless it fell within the conventionalist-positivist problematic and style of presentation. Yet then, it was likely to be obscured by other conventionalist preoccupations. Bridgman, Ellis and Bowman were noticed because they argued inside that problematic. But it is foreign to the real thrust of the definition. A structural issue was obscured by too much attention to Ockhamist, reductionist ambitions. There are further reasons for seeing this episode as underlining that our thinking about SR was partly misled by Einstein and Reichenbach. And seriously misled.

A blindness to issues other than parsimony and observationality obscured and confused the issue. It is not a simple matter of the scope of causal definition. Conventions for simultaneity can begin to make sense only in a particular – and imperfectly clear – conceptual setting: one where we speak seriously of the identification of places at different times and thus of the same time at different places. That is the language of frames of reference, not coordinate systems. Frames are distinct only if their rest points and their simultaneity classes differ,

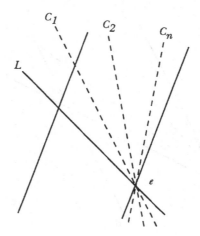

Fig. 1.2.

whereas coordinate systems differ more finely; for example, if we use polar rather than Cartesian coordinates for the same space of rest points. There is hardly an issue of convention or of anything else about simultaneity unless we take frames as somehow significantly different from coordinate systems. Frames of reference give us physics understood not through spacetime, but through space and time. The issue gets to be of significance only if we give some weight to taking space and time by themselves. There are no relevant non-factual, merely conventional statements in a spacetime treatment. Yet any convention in the one case should appear in the other. Relative to a coordinate system, the question whether or not events have different *time* coordinates has just the same status as the question whether they have different *spatial* ones. Neither question is seriously about rest or simultaneity. The space-like size of time-coordinate slices of things is well defined in coordinate spacetime physics, whether or not their spatial coordinates are identical for different time coordinates. But that is not the same as the thing's spatial shape and size. None of these mismatches can license a distinction between fact and convention as to simultaneity in a frame of reference of which there is no trace when we deal with coordinate systems. That is just one of several closely allied confusions.

Another is about the relation of frames of reference to coordinate systems for spacetime. That GR was written covariantly seemed to argue that it also lifts SR's restriction to the privileged set of Lorentz frames. But general covariance among coordinate systems is a much weaker condition than equality among frames of reference and complete symmetry for motion in all spacetimes – that depends on the structure of the spacetime itself. Clearly, inertial frames are preferred for flat spacetime. Someone might think that if skew coordinate systems are usable for a covariantly written theory, deviant relations of simultaneity in corresponding inertial frames must be permissible, too. But it is not so.

Second, the idea of a convention is the idea of a distinction between sentences which report facts and those which do not. It implies that real structure is less than we find in theoretical language. But the use of skew coordinates has nothing to do with that; it can only multiply the stock of things we related factual descriptions to. In fact, use of skew reference frames gives an unacceptable space which is anisotropic in all sorts of ways. Even if it did not, it would not give us a *convention.* It is not clear what role conventions can play in simultaneity. Sometimes the literature reads as if it is telling us that moving bodies have no definite shape and size in the frame. It seems to say that, though rest and motion are matters of relative fact, simultaneity has no such factual status. It waits to be determined, even after all relative matters are fixed. That is confusion.

The source of this lies in Einstein's treatment of simultaneity, which *seems* to license SR by purging some classical spatio-temporal structure. In fact, of course, it adds some. This could not have been understood before 1908 and the advent of spacetime. Certainly, SR removes the space stratification of classical spacetime, that is, its absolute partitioning into spaces at times. But SR replaces it with a metric of spacetime, whereas Newtonian spacetime, even if curved, has only affine structure. So SR has more structure, not less. Conventionalism really did obscure this and caused SR to be seriously misunderstood for quite a long time. There is another way to put this misunderstanding: it was not seen that the concept of reference frame, as we find it in classical physics, is inadequate for use in a spacetime with full metric structure. It needs revision. In classical physics, frame-relative descriptions of physical events are completely fixed by a choice of the spacetime curves to represent the frame's rest points. But SR is a more structured theory than classical physics, and choice of rest points does not give enough. Thus conventionalism told us precisely the opposite of the truth about the nature of this theory and its factual richness. It was false that a convention allowed use of familiar sentences, though a less structured world robbed them of factual import. Exactly the opposite is true; a factually richer world demanded a richer concept of reference frame to portray it fully. To describe all the richness of the world in frame-relative facts we need a reference frame which states its simultaneity classes as well as its rest points! The proviso about richness in frame relative facts is intended to cover, for example, the relative shape and size of moving things – their *spatial* characters rather than the *spacelike* dimensions of their coordinate cross sections. See also §4.

3 The relativity of motion

The relativity of motion is still imperfectly understood, I think. People thought it ought to get into physics as some sort of axiom, being either analytic, *a priori* true or, at the very least, self-evidently desirable. It is easy to see why; if motion is change of place, and place is a complex relation of distance and direction among bodies, then motion has to be a symmetrical relation among them. Other grounds were offered too, but epistemological demands on definitions for physical concepts have been among the more important. Many have felt that GR validates the relativity of motion in a way that meets the demands. So GR seems to commend, to scientists and philosophers alike, a positivist approach to spacetime foundations.

The relativity of motion is a big step toward getting space out of our ontology. That looks desirable because space is obnoxious to ontology and epistemology alike. It does enough work in classical mechanics to be indispensable but not enough to be intelligible. It seems featureless and intangible to the point of vacuity. Leibniz's objection, that it could make no difference to have created the universe somewhere else in space with all spatial relations among bodies just the same, looked plainly unanswerable.

But this was delusive. There is nothing *a priori* about the relativity of motion. Only naivety about geometry makes it seem so. It ought never to have figured as an *a priori* requirement on physical theory.

Still, something is wrong in Newtonian physics, and classical physics generally. It has been much discussed. Newton makes an absolute distinction between bodies that accelerate and those that do not. It is a causal distinction within dynamics. The related distinction between rest and uniform motion has no dynamical role; but Newton still wished to draw it. Later classical physicists spoke of acceleration absolutely, but of rest and motion only relative to some privileged frames of reference. They allowed themselves to speak of *changing* velocity absolutely without making sense of *velocity* absolutely. This is not satisfactory.

To call it unsatisfactory is not to call it false. The world could have been as Newton described it. It could also have been as the post-Newtonians described it. Each is unsatisfactory in a different way, and neither way is epistemological. Each description leaves the world incoherent; it gives us distinctions in kinematics which link with no distinctions in dynamics. It does not matter that identity of place is unobservable, but it does matter that it has no causal role in physics, while being a physical distinction. What is physics about if not space, time and matter? To put the ugliness of classical physics another way,

Galilean relativity gives us the sole classical example of change without cause; whether we see it as change of absolute place or change of distance and direction among bodies makes no difference. It is still an ugly affront to the classical presupposition that all changes are caused. The rest of classical science strongly vindicates this presupposition. The scandal of absolute space lies in that affront, not in its unobservability. But it gives a powerful motive for making the relativity of motion an *a priori* thesis.

However, rest and uniform motion are essentially observational concepts in a broadly classical setting; that is, one in which space and time are considered as separate. (If you prefer, think of a spacetime stratified at each time into a unique S_3 space, for instance.) This could not have been understood in the seventeenth century for no one knew about non-Euclidean geometries then. Spaces of constant curvature, either positive or negative, offer free mobility, as Helmholtz called it. That is, an elastic solid which fills any region of the space in a tension-free state, can so occupy any other. But an object which is in a relaxed state while at rest will not be relaxed while it is moving, if it moves in a constantly curved space. It will move under tension and distort its rest shape as a function of its speed and its deformability under a given force. Uniform motion of elastic solids is measurable by strain gauges in non-Euclidean space.

This is easy to visualise in the case of the uniform motion of particles in a dust cloud through a space of constant positive curvature. (I am assuming, here, a classical relation between space and time, not some sort of curvature of a spacetime.) For there, all geodesics eventually intersect and then diverge. In Euclidean space a dust cloud may keep its shape and size just as well in motion as at rest, since the velocity vector of each particle may be the same, in magnitude and direction, as that of every other; so the particles can move along parallel trajectories at the same speed. But this possibility is unique to Euclidean geometry, among spaces of constant curvature, at least. In elliptic space, the geodesics which the bits of dust move along intersect. The result is just as observable if the curvature is negative. We can consistently suppose that Newton's laws still hold for particles in such a space. A cloud of free-fall dust, in elliptic or hyperbolic space, can keep a volume constant in shape and size only if it stays at rest. Absolute rest is kinematically definable. Yet Newtonian laws apply to every point mass, so we here are envisaging worlds in which Galilean relativity still holds for them. The distinction has been made observable, yet still lacks a role in dynamics; *the breach of the causal principle is not filled.* This is still an incoherent world picture, ripe to create conceptual dissatisfaction.

Even in a world where we can never actually observe them, as in Newton's world, rest and uniform motion are in principle distinguishable by observation. The distinction passes rigid positivist and operationalist constraints. We can always raise the testable conjecture that space has some slight curvature to make practical the verifiability-in-principle of uniform motion. It was simply bad luck that the suggestion could not be made in the seventeenth century.

Variable space curvature yields stronger results. It cripples all those indiscernibility arguments which Leibniz used so persuasively. For in such spaces there is no isotropy and homogeneity from one point to another; spatial points differ in the structures that surround them, in purely spatial ways. So it really might make a difference if all the matter in the universe was one metre to the left of where it actually is. It could be that only that allows all matter to fit into volumes whose geometry lets them be relaxed there. There is no free mobility, even in Helmholtz's sense, if the curvature of space may vary. Note that I am not talking about GR here, where the indiscernibility arguments may be rescued. I mean a classical stratified structure, e.g. a constant S_3 at each time. In GR, the curvature of space accompanies the distribution of matter, so there is always another possible world like the one in question, but with all the matter and all the curvature displaced together in new regions. But if we imagine a possible world where curvature is not tied to matter distribution, this salvation fails (Sklar 1985, p. 14).

4 Non-decisive a priori criteria

Let us consider a Newtonian stratified spacetime, in which we make no appeal to curvature to account for gravitation. It offers a new perspective on the incoherence. This flat Newtonian spacetime is like a pile of hyperspatial sheets, pierced, as if with spokes, by time-extended particles. The sheets may slide across each other, just as a stack of paper sheets would do, taking the spokes with them and retaining the places where they pierce the individual sheets. Each paper sheet has its own

Euclidean spatial metric; the thickness of the sheets gives a time metric. The pile fills a Euclidean 4-space in which all force-free trajectories are straight. Nothing in the model corresponds to the length of a spoke (i.e. of a particle worldline). Nor is there a metric along the curved spokes of accelerating particles, of course. So nothing corresponds, either, to the orthogonality of any straight trajectory to the spacetime sheets they pierce. If we could speak of a spoke as piercing some *n* seconds of stacked sheets with a minimum spacetime length, then the spacetime would define orthogonal piercing and would define points of absolute rest and thus of absolute uniform motion. These concepts would be definite whether or not our perceptual powers equipped us to detect them. But though Newton presupposed the orthogonality, his physics fails to provide a metric basis for it. The affine structure of Newtonian spacetime has no such determinacy. Nor does it help to embed the pile of sheets in a curved spacetime where all free-fall particles are geodesics and gravity is caught up in geometry. So, again, it is incoherent, conceptually: the product structure of the spacetime yields both metric space and metric time, without any metric in the spacetime itself.

It is not inconceivable that the world should be this way, of course. But, I claim, this is *non-decisively* objectionable on *a priori*, conceptual grounds; that is, in such a world, scientists and philosophers would have legitimate grounds for seeking a more coherent conceptual structure, even if the world were just as this theory describes it. A true account of the world might leave us dissatisfied, with legitimate, specifiable, though not necessitating, grounds for trying to improve the account. In such a world there is a proper case for trying to reconceive it more coherently. We might try and fail, but not because we were obtuse. I am proposing *a priori* reasons which do not necessitate, but which appeal to criteria of coherence. This is something like the converse of Kripke's suggestion that there may be necessary propositions of identity which are not knowable *a priori*. Had Newton been right, I suspect that debates about space would have persisted unresolved in epistemology, physics and metaphysics. For, I think, space is not reducible to spatial relations among objects, yet its role in Newton's physics is obscure and unsatisfactory. That is, Newton described a quite possible, but imperfectly coherent, world.

I call these principles *a priori* because they lead theory choice; they are *reasons* for a choice. Principles of teleology, determinism, continuity and the principle that physics can be truly expressed in laws, are among them. They are not necessary, since teleology, determinism and continuity are either false or presently improbable. The principle that there are true laws is deeply entrenched, despite our belief that we do not yet know any law which is precisely correct. But it is not necessary. That is one way in which these principles are not decisive reasons for choosing theories.

I have tried to illustrate another. Whether such a principle is true or false may depend on the success or failure of some philosophical programme of reduction. I assume that either Newton's version or a post-Newtonian version of classical physics could be true, but not both. I also assume that not both a relationist and a substantialist version of these theories can be true. I do not think that a relationist reduction is possible, but a classical world would always present us with a good motive for trying to make it work, since it is an incoherent world: it disjoins uniform motion and mechanics. We still do not quite know how to settle these factual philosophical debates. Nor are we likely to find out until we learn more about how to identify and describe what I am here calling non-decisive *a priori* reasons round which the debate centres.

Principles of observation and coherence overlap. Any fundamental physical theory applies to every physical entity; thus every basic theory has the same domain as every other, though not the same ideology, of course. It follows that a theory will apply to measuring instruments and to observers and to processes within them involving the properties which the theory is about. The relation between the vocabulary of the theory and that of flashes, bangs, smells and tastes can only be as vague and unstructured as the latter vocubulary itself. This point hardly makes sense save within a realist context in which every subtheory in physics applies to a single domain of theoretical entities.

It may be, of course, that principles about the relation of scientific concepts to observational ones are among these *a priori* reasons which do not necessitate.

For these reasons, the supposed advantages of an unrestricted relativity of motion were illusory. On the one hand, rest and motion are, in principle, quite as observable and absolute as acceleration is. On the other, SR, in providing a metric for spacetime, provided the first metaphysically coherent and intelligible arena for physical events, in which, although absolute rest was not defined, its metaphysics gave no presumption that it should be well defined. It commits one, not to space and places, but to spacetime. Change of place is not a basic idea of the theory because space itself is not. The crucial distinction lies between linear and curved trajectories in spacetime. No way to identify places across time is needed since space and time, by themselves, have become mere shadows.

5 The geometry of force

Another thread woven into the fabric of the relativity of motion is an idea about force; that the spatial relation between the force's source and a test body which it acts on should be rather simple. Classically, forces meet something like the following criteria (Nerlich 1994, p. 177).

(a) Any force has an identifiable body as its source to which its target is spatially related in a definite way. Sources are force centres.

(b) The conditions under which the source body acts are specifiable independently of any description of its effect (e.g. a glass rod is electrically charged when rubbed with a silk cloth).

(c) Each centre of force acts on its target in some definite law-governed way (by contact, inversely proportional to the square of the distance).

(d) Generalisations describing the action of forces (e.g. all like-charged bodies repel) are defeasible (e.g. unless there is an insulating wall between them) but the defeating conditions must be causal.

Criterion (c) works in classical mechanics like this. In an inertial frame, any test body whose spatial path is not a straight line can be linked to a source object, so that any force vectors which curve the body's path, are related rather simply to the position vectors which point at the source. Even when the vectors of magnetic force round a moving electron are found to be orthogonal to the position vector linking a test body to the source, and in a plane orthogonal to the motion of the source, the relation is simple enough to let us see the field as an emanation of the source. For non-inertial frames, the force vectors for centrifugal and inertial forces have no such simple relation to sources. Indeed, that is plain from the stronger fact that the centrifugal and inertial force vectors will be the same whatever other objects there may or may not be elsewhere. But, I submit, it is only if the underlying geometry is Euclidean that the more general criterion linking force vectors geometrically to source position vectors is swallowed by the stronger criterion that real force vectors must depend *somehow* on the presence of sources. Only Euclidean geometry can realise the stronger, simpler criterion.

This use of criterion (*c*) is important for relationism and spatial reduction. That is because it gives a prominent, *physical* role to spatial relations among objects. The position vectors are spatial relations based firmly in dynamics; that plausibly contrasts with the wider embedding space which appears as a cloud of merely possible force and position vector links. I do not think this attraction overrides other difficulties which have been found in relationism, but I conjecture that the way it stresses spatial relations plays some role in reductive thinking.

There is some evidence of this in Mach's critique of Newton's rotating bucket experiment. Though Mach complained that Newton's theory conjures with unobservables, the real physics of his proposed answer to Newton's challenge makes the fixed stars sources of the inertial field. The field forces were envisaged as meeting the criteria just set out. Mach's proposal links centrifugal force vectors to position vectors for the stars in a way that was sufficiently simple, given the symmetry of the star shell. The relative motion of the symmetric shell through the non-inertial frame produces forces on the objects at rest in the frame. These force vectors at least *begin* to relate, feasibly, to symmetric position vectors and the direction of their change. The *a priori* complaint about unobservable space cannot be central, because the proposed revision results in more than conceptual changes. I am suggesting that Mach's real wish was to avoid the theoretical incoherencies that I am trying to illustrate and identify.

6 The relativity of motion in General Relativity

Whether the unrestricted relativity of motion is a thesis of GR in any philosophically interesting sense depends, as before, on our willingness to take rather seriously a distinction between frames of reference and coordinate systems. The distinction is, arguably, not clear enough to allow us to do so. If we want to study the physics of SR by the use of reference frames, we must note that 'accelerating' objects can no more function as frames of reference than they can in classical physics. But this says little about our willingness to work within the corresponding spacetime coordinate systems. So it is not perfectly clear that the sort of question that occupied the positivists about the relativity of motion is not simply left behind in GR. But I hope some light may still be shed on philosophical issues by treating frames in a serious, quasiclassical way. (See Jones 1980 for suggestions as to what counts as a frame in GR and for an excellent critique of the general relativism of motion.)

Why does the unrestricted relativity of motion fail for flat spacetimes? Firstly, because the very strong symmetries of flat spacetimes allow us to pick an infinitely large set of linear time-orthogonal coordinate systems, rather simply related to each other by the Lorentz transformation. To these there correspond rigid and global frames of reference, simply transformable into each other through relations which have the dimensions of a velocity. They are privileged frames in flat spacetime, because any other consistent frames of reference are either merely local, have anisotropic spaces, or have spatial and temporal metrics which change from place to place and time to time. So only a restricted relativity of motion holds here.

Nevertheless, though flatness reveals its strong flavour in this way, it produces Euclidean spaces (or everywhere locally Euclidean ones, e.g. in hypertoroid spaces) which, as I already mentioned, strike us as bland to the point of vacuity. It allows free mobility in Helmholtz's sense and it fails to distinguish kinematically between rest and uniform motion. So the very features which make flat spacetime yield a space which looks like a nothing, and thus ripe for reduction, also deliver an unequivocal preference for highly symmetrical and global reference frames. The preference restricts the relativity of motion, thus imperilling the reductionist programme.

We get the opposite state of affairs, in general, when we move to the more complex, variably curved spacetimes of GR. We may lose any or all of the criteria which make Lorentz frames desirable in flat spacetimes. To start with, there may be no possibility of global reference frames. But even where we can use them, we will find, in general, that they are not rigid; clocks will run at different rates in different places and space expand or contract as time goes on. That is to say, we will be obliged to treat the *coordinate* differences in time and distance seriously if we are taking the idea of frame-relative motion seriously. Of course, arbitrary choice of frame of reference may exaggerate this lack of rigidity wildly. But, in variably curved spacetimes, we lose a general contrast between rigid and non-rigid frames, and between frames that do and those that do not give us isotropic spaces.

Nevertheless, the same variable spacetime geometry, which swallows up preferred frames in its asymmetries, also, in general, gives the spaces of these arbitrary frames an obtrusive and changing geometry which has a very distinctive and prominent kinematic and dynamic character. It is impossible to think of a space as a featureless nothing, if voluminous solids are stressed, or even shattered, simply by moving inertially into regions where the geometry leaves no room for them to exist undistorted.

Consider the problem of an arbitrarily selected local frame in any spacetime, where we treat the space and time of the frame as given by the set of t-constant spacelike hypersurfaces and x_1-x_3-constant timelike lines that make up the coordinate lines and planes of some arbitrary coordinate system. The only requirement on the various, arbitrarily curved, space-like hyperplanes is that they must not intersect; similarly for the arbitrary time-like curves. When we project down from this arbitrary coordinate system into the space and time of the frame, the resulting arbitrary spatial geometry will have geodesics of its own which bear no simple relation to the geodesics of the spacetime (as it is easy to see from the special case of flat spacetimes). The motion of any free-fall particle whose spatial path in the frame is not geodesical, will be guided on its curve by vectors of the gravitational field, as determined by projection from the spacetime curvature of the region. These vectors can certainly not be expected to link, in any simple geometrical way, with position vectors, even of nearby sources of the matter tensor and the curvature guidance field, let alone of the array of sources as a whole. This will be especially true, of course, when the geometry of the frame's space is itself complex and changing. But even if it is not, we lose any simple link between force and position vectors. I have suggested that this gives most of its point to the focus on spatial relations, so prominent in reductionist literature. (See Jones 1980 for a rather different perspective on what relativity of motion actually calls for and for a searching critique of it.)

Finally, GR is rather rich in spacetimes which yield preferred reference frames (or classes of frames), both globally and locally. Nor are such examples among the rarer and more exotic models. Flat spacetimes are examples, as is the spacetime model of our universe preferred in modern cosmology. In fact Rosen has suggested, in several papers (e.g. 1980), that GR might be rewritten in such a way as to make this fact formally more prominent. This has not been widely accepted, but it serves to emphasise the unintended emergence of preferred frames.

It remains true that the basic concepts of GR are relativistically forged in that the theory is written covariantly in a non-trivial way. But that does not mean that the unrestricted relativity of motion is a key concept of the theory, nor that the best way for us to learn about the metaphysics of spacetime structure is to understand the mistakes of the positivists or the philosophical relativists more widely. Their ideas were often confused and baseless.

We might ask whether the nature of the question changes if we approach it in a way less tied to classical and to positivistic ideas about how to answer it. That does give rise to a rather different kind of discussion, but not one which suggests that the positivists were on to something after all. Positivists are strongly inclined to think, like relationists generally, that spacetime relations ought to be analysed as somehow or other causal ones. But in terms of what is generally meant by causality in modern treatments of spacetime (a meaning strongly influenced by positivists themselves) motion is clearly not generally relative in the spacetimes of general relativity. Malament (1985) shows that non-rotation is an invariant property of conformal transformations. The conformal structure of spacetime is essentially the null cone or light cone structure and regarded by positivists as strongly causal. But then the causal structure of spacetime determines the distinction between rotation and lack of it in an absolute fashion. Roger Jones (1980) penetratingly considers a variety of suggestions as to what should constitute a general relativity of motion. He shows clearly that GR is rather restrictive in that regard.

7 Other ways of understanding spacetime

Some think that only conventionalism and positivism offer any systematic picture of the foundations of spacetime theories (Sklar 1985,

p. 303). But this is not so. Conventionalism avowedly fails to find any foundation at all for large parts of the theories, writing them off as mere non-factual conventions. Where it does offer foundations they are narrow, opaque or dubious, as in the case of causality, or confused as in the case of simultaneity. A more pluralistic view of the search for foundations is needed. We need a metaphor of foundations without foundational*ism*.

Let me turn now to describe three ways, quite different from each other, in which we might properly regard a philosophical study (or a quasi-philosophical one) as tracing the foundations of spacetime theory. These by no means exhaust the range of options and are not even intended to represent the traditional main stream of what might be called foundational studies.

Positivist and conventionalist attempts to find the foundations of physical theories are motivated, clearly enough, by the laudably modest aim of saying no more than one must. Yet they also have an immodest tendency to rob theories of legitimate assertive power.

A taste for modesty may lead us to substitute a particular, favoured expression for other extensionally equivalent but intensionally different expressions. Relativity theories provide several striking examples of this, most notably, the very common reference to the null cone as the light cone. The world 'causal' is very widely used for the relation, whatever it is, that divides the surface and interior of the cone from what lies beyond it. But, where positivism hardens its heart against intensional distinctions among coextensive expressions, its drive for economy is no longer simply modest. This is clear when one contrasts it with a modesty in realist attempts to arrive at a foundation for theory. I turn to that in a moment.

In general, positivist investigations search for what I shall call the *restrictive* foundation of theories. That consists of a minimal ideology – foundational*ism* in short – and a minimal set of axioms, which generate a body of theorems previously judged as indispensable. To call this the restrictive foundation stresses its tendency to reduce content and prune ideology. This is a bald account of positivist aims, to be sure, but I hope their familiarity will allow me to be brief and turn to something less familiar.

One motive behind a realist examination of a theory, is to look for what I shall call its *permissive* foundations. Whereas the restrictive foundation presents, in few axioms and a lean stock of predicates, all the theorems we simply must have, the permissive foundation presents axioms and predicates within which we can speculate most radically on how that same theory may develop. We can reflect on the results of dropping quite deeply entrenched propositions from a theory. Radical speculation can legitimately rest on the theory's broader base for its development. The permissive foundations tell us what this base is, and which theories might yield to correction without the theory's being abandoned. It is not clear that these restrictive and permissive motives for finding foundations must lead in different directions, but in space-time theory they certainly do.

It is very widely believed that a single principle about causality lies at the foundations of the relativity theories. It is the Limit Principle, that nothing outstrips light. The nice things about this Principle are that it is, very likely, true, that it is qualitative, and expressed in lean and intuitive ideology – simply 'particle' and 'outstrips'. If we take this as giving the core of the theory, we are likely to include it among the axioms, and to allow ourselves to speak of the null cone as the light cone, since photons will lie in the conical surface. That is its restrictive foundation.

But there is a price for this. It may cripple speculation within the theory. Questions which might be fruitful cannot be pursued as developments of *that* theory, nor call on its resources, to make speculation clear and definite. The hypotheses that there are tachyons, and that the photon has some finite mass, cannot be pursued within SR if we formalise it according to its restrictive foundation, yet there is a clear sense in which both these suggestions have been investigated consistently within SR (see Feinberg 1967 and Goldhaber and Nieto 1971). Unless each was compatible with the foundations of SR, in some clear sense, neither could have been considered. The idea of permissive foundations for a theory may give to realism a theoretical modesty of its own; for it is clear that there may be a proper diffidence about denying structure just as there is about asserting it. In the present case, suppose that we admit null cones, which are observationally remote structures in an equally remote object, spacetime; then a realistic attitude towards them lets us speculate about epistemically more proximate objects, tachyons and photons, in a tangible and articulate form given by the admission.

When we ask for a permissive foundation, the question is quite differently motivated, and our criteria for a good answer are not at all the same, as for a restrictive foundation. Though the two hypotheses just mentioned are improbable, we can usefully ask how SR would survive their truth. The answer is quite obvious from a glance at tachyon theory, for all of its equations are Lorentz invariant. So the permissive foundation of SR is the Invariance Principle, not the Limit Principle. The Invariance Principle simply refers to the conical structure within spacetime and to the very powerful symmetries of Minkowski spacetime. The hypothesis of the massive photon may call for some changes in the laws of electromagnetism and optics, but whatever laws are suggested are still required to be Lorentz invariant. So that Principle is the permissive core of SR.

Third and lastly, I want to discuss Ehlers, Pirani and Schild's well-known paper (1972) on the foundations of GR. In this, the authors link various geometrical structures – projective, conformal, affine and so on – material structures such as photon and particle free-fall trajectories. In what way is this paper foundational?

Two physical conditions are needed for it to work at all; particles in free fall and null cone surfaces filled by light. However, GR is not a theory about the constitution of matter and has no ontic commitment to matter in particulate form. It can admit tachyons, and speculation about massive photons. For these reasons, we cannot see Ehlers, Pirani and Schild as presenting an ontology, nor even a likely epistemology, in any standard sense, since neither particles nor the geometry of light propagation are observationally proximate. Yet something like each of these studies is at stake in the paper.

Free-fall *particles* are needed because only point masses can be relied on to trace out the geodesics of spacetime which fall inside the null cone. These trajectories constitute the projective structure of spacetime. The centres of gravity of voluminous, elastic, massive bodies will not do, since the geodesics through various points in the solid will usually not be parallel. The bodies are gravitational multipoles. Internal stresses will tend to force the falling body off geodesical trajectories, even in the absence of external forces. The worldline of neither its geometric centre, nor its centre of gravity will be geodesic. So, in general, it is only freely falling *particles* which inscribe geodesics. Similar reflections apply to the filling of the null cone. Of course the surface of the null cone will still be the boundary between space-like and time-like curves; it will be the overlap of points elsewhere and elsewhen from the apex of the cone whether or not light or matter fills it. So what Ehlers, Pirani and Schild offer us is an elegant and familiar way in which the geometric structure of GR might be *inscribed*. They single out a subset of worlds, which the theory makes possible, in which the geometric structures at the core of the theory and matter structures, familiar from a range of other theories, come together so that the latter trace out the former. This might be called an investigation of the inscriptional foundations of a theory, presenting one possibility within it.

The idea of an inscriptional foundation suggests an *ideal* way in which our understanding of the structure of spacetime might be traced out by matter. Real matter is not required by the theory to lie in the suggested trajectories. Recent work by Coleman and Korté (1982, 1993) argues that it may be rather less an idealisation than I have just suggested. They argue in detail, first that we can identify paths which are geodesics (and thus the projective structure) of spacetime in terms of a criterion framed in the language just of differential topology. They show how to trace particles from any coordinate system which we can identify by the smoothness and continuity of its curves and to verify whether these particles then follow the geodesics defined. They show how to correct for the effect of forces on charged particles. This, allied to the work just discussed, allows us to show clearly how both projective and conformal structures may be inscribed on spacetime in such a way as to define a Weyl structure yielding an affine connection and, eventually, a metric for spacetime.

I close with the suggestion that the philosophical investigation of physical theories may take several useful forms, that the form of a theory that philosophers find in scientific use is likely to be irreducible, and that time spent in attempting to reduce it to something else may illuminate why the theory is composed as it is, but will seldom result in a justifiable revision of it. Of course investigations into the epistemology of a theory shed light on what goes on in it, but epistemic theories of concept formation have seldom proved constructive or even insightful. One can usefully recognise *a priori* elements in a theory without relegating them to mere convention, and hope, eventually, to see how they are corrigible by observation. In short, there is a variety of ways in which we might look for foundations in physical theory, for many of which it is simply unhelpful to complain that they raise epistemological problems which they make no offer to solve.[1]

1 I am grateful to Graham Hall (Department of Mathematics, University of Aberdeen), Adrian Heathcote (University of Sydney) and Chris Mortensen (University of Adelaide), all of whom read the paper in some form and discussed it with me. They are not responsible for any mistakes.

2 What ontology can be about: a spacetime example

With Andrew Westwell-Roper

At least since Quine's seminal work *Word and Object*, ontology has been dominated by a particular model of ontic commitment, that a theory commits itself to entities through the variables of quantification and what values those variables must take if the theory is to be true. 'To be is to be the value of a variable.' The predicates of a theory do not carry the load of any commitment to what is in the world itself.

The metaphysics of spacetime does not lend itself elegantly to that style of analysis of its picture of the world. Of course there is the central ontic question whether modern physics is committed to an entity, spacetime, or whether this can be reduced or 'parsed away' in a restructured theory. But so much of the metaphysics of spacetime concerns the various levels of structure which it has – whether it is merely a differential manifold, whether it has projective, affine or metrical structure, at what level its structure may cease to be objective and give way to merely conventional determination. This seems likely to bring us to consider some sense in which the predicates of theories rather than their quantificational structure carries some form of ontic consequence.

In this essay, jointly authored with Andrew Westwell-Roper, we explore ontology of structure rather than of quantification and the kind of consequences for methodology in metaphysics is probed. We touch on the work on the foundations of special relativity due to A. Robb and John Winnie

1 Introduction

Quine's original idea about ontology was that ontic questions are referential ones: they ask what *objects* there are. Our best answers to ontic questions tell us what our best theory, in its canonical form, says that there is. We get the answer from the theory by finding what objects must lie in the value range of its quantified variables. So ontology is an objectual matter.

This position evolved somewhat when Quine (1976) argued that *all* the objects quantified over by *any* physical theory could be reduced to nothing but pure sets, although the theory applies non-set-theoretical predicates to those sets. For example, the predicate 'is an electron' might be applied to the set of coordinate quadruples for what were, before the reduction, the spacetime points in the four-dimensional region occupied by an electron. Hence, it was suggested, the ontological question of what objects must lie in the range of the quantified variables is no longer of interest, since those objects can always be pure sets. Rather, the ideological question of what predicates are required by the canonical form of our best physical theory 'is where the metaphysical action is'.

We want to pursue an idea about ontology that would complement Quine's. The objectual questions with which Quine has been concerned are now to be cast as questions about which *monadic* predicates the lexicon of natural science must contain; must it contain the predicates 'is an electron' and 'is a particle', for example? We wish to focus on the polyadic predicates required by natural science, and the *structural* questions they generate. These are questions about the kinds of relations in which the objects of this world stand. (We are not concerned here with *objectual* questions about about whether relations exist over and above their relata.) Quine's metaphysics does not tell us much about how to find the structures which the world must have if a given theory of it is a good one, but we contend that an examination of the structural commitment of a given scientific theory can lay bare certain objectual commitments which are not discernible from its quantificational apparatus and monadic predicates.

Philosophical work on spacetime abounds in clear and crisp examples of argument about ontic structure. Our tactic will be to present some of these as briefly and simply as possible in a way that shows the point of treating them as ontic questions which are not objectual ones. Quine's view of philosophical argument finds a different place for these discussions – the place of how to write a theory in its best canonical form. But we think that this obscures what is most interesting about them. They concern how far one structure of the world may underlie certain other structures as their basis, and they concern just what this basic structure really is. We argue, at the end of the paper, that the answer to the main objectual question of spacetime physics – whether spacetime itself is in the ontology of our best theory – depends on answers to the structural questions.

The main structural questions of spacetime are standardly expressed in this form: can we identify F structure with causal structure? Here F dummies for 'topological', 'conformal', 'metric' and other structural descriptions. Something is wrong with this standard form. The trouble is with 'causal' (as a number of writers agree). We will say why that is not quite right at some length, later. Elsewhere, one of us suggested that the question was about what structures we could get out of the *topology* of spacetime. That is a little misleading, too, for we could obviously get topological structure from topology. Perhaps the flavour of these structural questions might best be given to the general philosopher by phrasing the main question (about metrics) thus: can we identify the quantitative structure of spacetime with a qualitative one?

Conventionalist theories of spacetime give a positive answer to the question, but only in a weak sense. In these theories, all spacetime structure is causal, but this only means that there is, in fact, no metrical structure in spacetime itself. Our theoretical *language* has a metric structure but this is really all talk, so the conventionalist tells us, and which sentences of this language turn out true rests on arbitrarily chosen conventions and reflects no structure in the world. We will not talk about conventionalism in this paper.

According to the theories we want to discuss, the metrical structure of spacetime may be fixed uniquely by its causal (rather, by *some* qualitative) structure. There is no question here of metrical language as something like superfluous baggage. Even in the examples where the programme fails one cannot equate the metric with causality by robbing metrical language of its apparent content. Either causality identifies the complete unique metric congruence structure (the topology, or whatever is in question) or the project fails.

These theories are about structures in spacetime, not about objects in it. They tell us what relations hold among these structures and which are primary or which independent. They answer questions which are recognisably about scientific realism, in one of its forms, and what is that if not ontology?

2 Structures in spacetime

In this section we want to report some results very briefly: successes, failures and prospects. In the next, we comment, equally briskly, on doubts about whether the basic relation really is causality. Later sections get back to some philosophical repercussions of the results and the doubts.

As long ago as 1911 A. A. Robb completed a remarkable account of Minkowski spacetime which treated its whole metric congruence structure in a coordinate free way and which required no use of rods or of clocks (Robb 1914). The theory of these general classes of measuring instruments standardly involves quite complex quantum mechanical ideas to explain their (rough) invariance. This seems unsuitable in the foundations of a theory which need lay down no quantum postulates for its own main development. Robb's treatment takes a single qualitative relation, which he called *after*, and postulates a number of conditions for it. These turn out to be enough to determine all the geometric structure for Minkowski spacetime which it seems reasonable to ask for. Consider these examples.

(a) Select a time-like line in Minkowski spacetime and choose a point on it. Then one can define a causal structure which picks out a hypersurface orthogonal to the line at the point. It is defined (roughly) by the intersection of two light cones from two point events on the line such that the given point is midway between them. The significance of the construction is that orthogonality is thus tied to simultaneity in a reference frame in a direct, not a conventional way; for the (hyperplane) of events orthogonal to the line at the point is just the class of events simultaneous with the point event on the line (with respect to a frame for which the line represents at rest point). The simultaneity-classes definable in this way are just the events simultaneous by standard signal synchrony, those made simultaneous by setting $\varepsilon = \frac{1}{2}$ in short, just the simultaneity classes we want. It has been shown that more complex, less 'natural' simultaneity classes proposed as equally acceptable by conventionalists cannot be defined by any other purely causal construction. Thus the *standard* sets of simultaneous events in any reference frame can be selected without arbitrariness by a purely causal construction (see Malament 1977b).

(b) A series of causal definitions of curves, lines and planes leads on to the whole metric congruence structure of Minkowski spacetime. For any pair of intervals one can find a construction which identifies them as either congruent or not, whether the intervals be spacelike, time-like or light-like. Much of the work lies in showing that the construction, based on causal connection or *after*, does actually correspond with the intended geometrical object. Thus it is by a series of definitions that the causal or 'after' vocabulary is, in certain constructions, *coextensive* with the geometric one. It shows that we can identify metrical classes with qualitatively specified ones.

This is a surprising and, perhaps, a gratifying result. It appears to carry out Leibniz's causal programme in a rigorous way, for all spatial and temporal quantities and relations are, indeed, identifiable by means of causal conditions, so long as one sticks to Minkowski spacetime. We can grasp its structure without reliance on the methods of coordinate systems or reference frames and without the use of instruments (rods or clocks) extrinsic to the structure we are probing. In the end, it looks as if we can reduce all this to a matter of concrete relations among concrete events (Winnie 1977).

Unhappily, the result does not easily extend from the case of Minkowski spacetime to Riemannian spacetimes generally. The homogeneity of Minkowski spacetime allows causal constructions to incorporate global constraints on which the success of Robbian definitions depends. But in the full range of spacetimes envisaged in General Relativity one cannot even define the topology of spacetime, in many cases, by means of Robbian constructions. Clearly, this fact weakens the case for saying, even for Minkowski spacetime, that the metric is a causal structure rather than merely that it *coincides* with one (Sklar 1977b).

Nevertheless, the idea that causality – that is to say, the idea that some basic qualitative relation rather like causality – is of fundamental physical importance is strong and apparently getting stronger. Thus Robert Geroch (1971) claims a widespread belief among relativity theorists that the first, and most basic conditions to consider in any example of a spacetime, are the causal ones.

The situation in the literature, which is by now vast and almost all of it very complex, is a little strange. Exact solutions of the Einstein equations are not easily had and, as we understand it, there is still no synoptic view of the range of examples permitted. Furthermore, it appears that a wide range of spacetime structures that are quite unrealistic, physically, are worth studying, both because they are more easily surveyed in a more or less purely geometrical way, and because some form of spacetime theory may well survive the possible refutation of the Einstein equations. Causal theory seems to be studied as a means of probing not only the physics of GR but also geometric spacetime structures more widely. It is a way of studying what generates the geometry, perhaps in the hope of showing that something physical does (Geroch and Horowitz 1979). We will discuss this hope later.

(c) Last, it has been shown (Malament 1977a) that if we enrich the basic causal structure in an intuitive way then we can get at least the topology of any kind of spacetime out of the enriched structure. The basis is causal *curves*. It can be shown that every transformation of a spacetime which preserves its causal curves preserves all of its topological structure.

3 Is causality really causal?

We have been hinting rather heavily that, although the literature abounds with references to causality and the study is called causal theory, it may not really be causality that is at issue. Some qualitative relation is widely felt to generate spacetime structure, at least locally. Whatever the relation is, it must serve to pick out the class of points inside or on the surface of the null cone at a given point. That is obviously a formal requirement on the constructions causal theorists build. Causality is simply the name for whatever qualitative relation meets this requirement. But some uneasiness is often expressed that this relation may not actually be causality.

A reductive hope is enshrined in the expression 'causality'. It is the hope that Leibniz's programme might succeed, or that something like it may. Perhaps geometric structure can be quite strictly generated out of material structures. If so, that would be significant for the objectual ontological question whether spacetime is a thing which exists independently of material objects and material relations among them.

There are several grounds for thinking that it is seriously misleading to call this fundamental relation causality. First *actual* causal connections are clearly not adequate to define what is wanted. We would have to include causal connectibility: pairs of events which *could* be causally connected. That no longer sounds like a purely physical relation. There are familiar worries about the use of modal concepts in reductive analysis which need not be rehearsed. Perhaps we should voice a conviction that the use of modalities is standardly a sign that the alleged reduction is actually circular. This case can be made out with particular clarity and certainty in many instances within spacetime metaphysics (Nerlich 1994, pp. 65–67). We take it to apply quite generally in the field.

There are two versions of the non-modal idea of causal connection. First, events are causally connected if they are linked by a particle, either massive or massless. Alternatively two events are causally connected if a signal is sent from one to the other. There are several things wrong with these suggestions but we will not mention them all (see §3). Some comments will make things clearer about what kind of study the ontology of structure is.

(a) At the end of his paper (1977, p. 198) on causal construction John Winnie points out that the basic relation, whatever it is, has to be instantiated in all spacetimes which are solutions for the equations of GR. This includes the so-called vacuum solutions in which the world is empty of matter. In these solutions mass and energy are associated with gravitational waves and with varying and evolving curvatures of spacetime. So something goes on in these universes, matterless though they may be. No particle connections occur in them, however, nor is there any electromagnetic (or indeed any non-gravitational) field through which a light signal, or any other kind of signal, could be sent. Hence there are no causal connections according to either of the above two views of causal connection. Though we might hope to speak here of gravitational influences and, therefore, causal ones between one spacetime point and another, it is not clear how this avoids presupposing that the very structures one hopes to probe are curvatures (ripples etc.) in spacetime.

This objection to causality plainly does not rest on its not having the right *extension* in the actual universe, as a matter of fact. The point is, rather, that there are models where causal connectibility (causality) rests on no basis of actual causal connection in the model universe. In our world, at least connectibility has a non-modal basis. But the empty spacetime solutions leave the link between the qualitative relation and causal connection unsatisfactorily attenuated.

(b) Several years ago, a number of papers appeared in physics journals raising the question whether or not there might be particles, to be named tachyons, that go faster than light (see e.g. Bilaniuk and Sudarshan 1969; Recami and Mignani 1974). The question whether Special Relativity could consistently incorporate these fast particles was briskly debated but no clear inconsistency was shown. If tachyons are consistent with SR, then they are consistent with GR. No such particle was ever found, though some experimenters thought they had detected them in results that were not replicated. Discussion lapsed somewhat when it was found that the quantum fields of these particles were not stable. The hypothesis is inert, at present, but hardly dead. A revival has been mooted by Michael Redhead (Heywood and Redhead 1983), for it may help to ease the more painful aspects of recent versions of the EPR paradox in quantum theory. But what matters for us, here and now, is that the hypothesis that there are tachyons is not clearly inconsistent with special relativity. It is, however, *quite flatly inconsistent with the claim that causal connectibility is coextensive with the relation we want.* The tachyon hypothesis just is that particle connections fall outside the null cone. It plainly follows from the hypothesis that particle connections would not serve to identify the surface and interior of the null cone. This shows conclusively that there is no *fundamental* postulate of relativity which says that nothing goes faster than light. If there were, one could deduce the inconsistency of tachyons with SR in a one-line proof.

(c) Over the last twenty years there has been speculation among physicists over whether the photon has non-zero rest mass. According to the standard formulation of quantum electrodynamics (which is both a relativistic and a quantum theory), the rest mass of the photon is zero, so that, as a consequence of the relativistic equations relating the mass, energy, momentum and velocity of a particle, its speed is always the frame invariant constant c. However, other formulations have been proposed which retain SR as the relativistic basis of quantum electrodynamics, but reformulate the electrodynamics so that the rest mass of the photon is a constant the value of which is open to experimental determination (Goldhaber and Nieto 1971; Barnes 1979, §1). If that constant were to have a value greater than zero, then the speed of the photon would be less than c, i.e. light would not travel at a frame-invariant speed in the null cones.

This may seem puzzling since the invariant speed of light plays such a prominent role in most expositions of SR, and it may well seem that to abandon this simply is to abandon SR. It is argued later that this conclusion does not follow. For the moment, we simply record the claims that SR is consistent with light's lying in the interior of the null cone, never on its surface.

Those who postulate a non-zero photonic rest mass sometimes maintain that even if light does not occupy the null cones, nevertheless some other causal process does. Thus Goldhaber and Nieto (1971) suggested that the neutrino, which at that time was thought to be massless, provides the basis of spacetime structure. The neutrino has since been found experimentally to have a small positive rest mass, but the graviton is still thought to be massless.

There are two reasons why the reductionist cannot resort to identifying null cones by means of possible graviton trajectories, however. First, our concern, like that of the reductionist, is with the ontologies of SR and GR, not of quantum electrodynamics or quantum gravity. Neither SR nor GR describes either the photon or the graviton, and we cannot provide ontological foundations for a physical theory in terms of objects which the theory does not explicitly recognize, and could not even implicitly recognize, since it could not describe them adequately. Our discussion of the photon was intended to show only that the theory that the speed of light (conceived as a classical electromagnetic wave) is less that c, might be thought compatible with SR. (The reformulation of quantum electrodynamics which allows for a speed of light less than c involves only a deviation from the classical equations of Maxwell, and does not depend on any quantum aspect of electromagnetism; cf. Barnes 1979, §1.) Similarly, for our purposes the zero rest mass of the graviton in quantum gravity just reflects the fact that the gravitational waves which are solutions of the GR field equation travel at c.

Secondly, and more importantly, the fact that gravitational waves in GR travel at c cannot be used as the basis for identifying the null cones in terms of causal processes in SR or GR. It cannot be so used in SR because SR does not recognize and cannot describe a gravitational field or waves in such a field. It cannot be so used in GR because there gravitational waves are reduced to fluctuations in space-time structure. Thus, in order to identify a gravitational wave one must first identify the spacetime structure, so that gravitational waves cannot be used to identify some aspect of spacetime structure, such as the null cones. Of course, the reductionist claims that, in one sense, gravitational waves are *not* fluctuations in spacetime structure; he will maintain that spacetime structures are ideal and can be reduced to causal processes of some kind between material bodies. However, to effect this reduction he must be able to reduce at least the conformal structure of the null cones to some system of causal processes, before he can hope to reduce the full metrical structure, including the fluctuations in spacetime curvature involved in gravitational waves, to a system of causal processes. That is, the reduction of gravitational waves to causal processes presupposes the very capacity to identify the null cones which the reductionist is trying to achieve.

(d) Sklar (1977a) suggests that the best reason for an interest in causality lies in the causal curves. That is, curves which are, everywhere, time-like. He claims an epistemic advantage for these. It is that each curve represents what might be the worldline of a consciousness. Seen in that light, it is not the causality of the curve that counts, but rather the fact that we have basic experience of ordering the events which lie on not too large segments of such curves. So Sklar thinks that causality is a misnomer. We have an intuitive understanding of what counts as events being close to one another on such a curve. Once again, it seems questionable whether the basic qualitative relation we want to identify is causality.

4 What is ontic structure?

From these examples it would seem that the problems of ontic structure are not extensional ones. It seems likely that causal connectibility and the basic qualitative relation between any point event and any other on or in its null cone are coextensive. If there were tachyons, or if photons had non-zero mass, the relations would not be coextensive, but there probably are no tachyons or massive photons. The questions raised in (b) and (c) of §2.3 concerned only the *consistency* of these hypotheses with relativity theory.

If structural questions are not extensional, then perhaps they are intensional, i.e. questions about which relations-in-intension must be the meanings of those polyadic predicates required by a given scientific theory. However, the attempt to reduce spacetime to a system of relations of causal connectibility discounts this suggestion. For this is an attempt to *identify* causal connectibility with the basic qualitative relation. While there are problems surrounding the identity conditions for relations-in-tension, we may take it that such relations are identical only if they are (logically) necessarily coextensive, and hence only ever identical necessarily. Causal connectibility and the basic qualitative relation clearly are not *necessarily* coextensive, however; the reductionist can only be claiming that these relations are *contingently* identical, so he cannot be talking about relations-in-intension.

Relations which are not relations-in-extension, and admit of contingent identity, should be treated as physical relations instantiated in our physical world, not as linguistic meanings. Their identity conditions may not be completely straightforward (Putnam 1969) but they are not the intensional entities which Quine has rejected for their lack of identity conditions. For our purposes we need only set a necessary condition for the identity of physical relations, viz. that physical relations are identical only if it is *physically* necessary that they be coextensive, where what is physically possible and necessary is determined by what is consistent with the physical theory at hand. The consistency of various non-causal hypotheses with SR or GR is sufficient to show that causal connectibility and the basic qualitative relation are not the same physical relation.

Another structural problem can be put, a bit roughly, like this: we want to order the ideology of the theory or at least its polyadic predicates as any formal system orders its ideology; so that some relations are seen as primitives, others defined, and so on. That ordering tells us which relations the formaliser thinks lie at which levels of basicness.

In dealing with this problem in spacetime we have a huge advantage of precision in that we are dealing with structure *in geometry*. Various levels of structure can be precisely defined and ordered. There are, for example, topological, projective, affine and metric structures in decreasing order of generality. One can define the levels by the group of transformations of space into itself which preserve structure at these levels. For instance, topology studies those properties of space which are unchanged when it is arbitrarily (but continuously) distorted: these are the properties of separation, neighbourhood, boundary, enclosure, dimension. Roughly, projective geometry studies properties which are preserved when space is distorted in any way which keeps lines straight. Thus each geometric level of properties is precisely defined by the group of transformations which preserve the properties of that level. This means that we can order geometric properties, as to which are the more general and deep-seated, in a very satisfactory way. As a consequence, the constructions on the qualitative relation which aim at capturing topology, affinity or the metric either succeed exactly, or they fail.

Looking for ontic structure involves pondering the ways of some possible worlds. One considers models of GR which are vacuum solutions of the Einstein equation, or models in which the photon is massive or in which there are tachyons. We suggest that this does not befog our structural question because several of the difficulties which cramp the style of possible world metaphysics in general are missing here. First, the candidacy of a model rests on the necessary condition that it be a solution to some well-defined set of equations. Second, we need not be concerned to fix what constitutes the range of all worlds accessible in that they are solutions to the equations. It is frequently enough to look at a few salient or, perhaps, limiting cases to settle main issues. Lastly, there is no evident need to raise questions about the identity of objects in one model with objects of another. We can forget the problem of Trans World Heir Lines (to borrow David Kaplan's engaging phrase).

For all these reasons, then, it seems that the study of ontic structure, need not be less determinate than the study of objectual ontology. It is rather that some physical relations are less tractable than others and that hope in settling questions about physical relations must be tempered by one's sense of how fully the target theory has been articulated.

5 Studying structure

If one branch of ontology is structural rather than objectual, what is the best way of studying it? The discussion so far suggests that we should formalise our theory and use the quite well-understood resources of formal systems. Choosing axioms, definitions and theorems helps to present just the sort of map we want of a theory's ideology. The suggestion has already been pursued in detail by Sklar (1977a). We add some observations which may help to clarify and support our claims.

(a) Positivistic probes. Positivist and conventionalist critiques of theories are familiar. They are motivated by the fact that two mutually inconsistent theories may each entail the same body of (actual and possible) observation statements. Theories may be underdetermined by observation, in short. So, in the construction and axiomatising of theories, we face observationally unanswerable questions about which theory to choose. The positivist strategy rejects the questions either by pruning away the vocabulary in which they might be asked, or by so defining the vocabulary in observational language as to make the answers analytically true or false. These recipes tend to yield rather strong medicine. Less stringently, the conventionalist strategy rejects the question by robbing some part of the ideology of its fact- or structure-reporting role. The question gets an answer, though not a structurally significant one. These three styles of treating the problem of underdetermination are motivated by laudably modest wishes to say no more than one can confirm. However, the characteristic fault of such critiques is their strong tendency to evacuate the theory of its assertive power. Its content implodes under positivist pressure.

The somehow laudable distaste for excessive content may lead not just to the three kinds of ideological pruning just mentioned. It also promotes a tendency to substitute for one another extensionally identical expressions which are not intensionally identical. It does not require that, of course, but relativity theory provides clear examples where it has been done: notably, the almost universal reference to the null cone as the light cone. Use of the expression 'causal' seems to have been adopted, also, in the hope that it would be coextensive with the qualitative relation we are pursuing, so as to offer a reduction. In as much as positivism hardens its heart against intensional distinctions among coextensive expressions, its drive for economy ceases to be powered by a seemly modesty. This is clear once one sees a certain modesty in realist aims as well. We will turn to that in a moment.

In whatever way they work, positivist probes seek to discover the core of spacetime (or whatever other) theory. The core consists of the minimal ideology and the minimal axioms which use it, so long as they suffice to generate the needed theorems. Thus positivist probes seek what may be called the *restrictive core* of the theory, thus stressing its tendency to reduce content and prune ideology. This is a rather bald account of positivist aims, but we hope the familiarity of the ideas touched upon will allow us to be thus brief.

(b) Realistic probes. Motives which reflect the scientific realism of theorists in choosing formal systems may vary. Clearly, one motive must be to avoid the implosion of theoretical content which positivism is so likely to produce. Among other motives, we wish to stress two which are well illustrated in spacetime metaphysics.

The first is an argument addressed specifically to Reichenbach's (1958) treatment of conventionalism in geometry. Reichenbach argues that two theories underdetermined by the same set of observations have the same factual content. But Clark Glymour (1977) has argued against this. Though the set of observations relevant to some pair of inconsistent theories may be exactly the same, so that they are empirically equivalent in that sense, this body of evidence may still confirm one of these theories better than it confirms the other. This is to say that the theories are distinct and non-synonymous. It is also to say, of course, that our choice between them is not arbitrary. To show this is a matter of articulating in spacetime examples a certain schematic principle of confirmation. It says that a theory with fewer untested hypotheses is better tested by the same body of evidence than a theory with more untested ones. More determinately, let there be two theories P and Q, a set of definitional axioms A, such that P & A entail Q, but no set of definitional axioms B such that Q & B entail P. If every test of P & A from some body of evidence is just a test of some hypothesis in Q, then every body of evidence confirms Q better than it confirms P. We will not pursue this in more detail now.

Glymour's result, whether it is right or wrong, points to something important in motives for scientific realism. One great weakness of positivist critiques of physical theory surely lies in the assumption that we understand epistemic principles better than we do physical ones. But, surely, we do not. It is entirely reasonable to regard the failure of modern epistemology to provide an adequate rationale for the ways we choose our theories as demonstrating the incompleteness of epistemology. At least, that strikes us as quite clear. It is always open for a realist to say that the current state of epistemology shows, not that science is in bad shape, but rather that some relation between simplicity and truth or some new theorem in confirmation theory such as Glymour's, is a good bet for a research programme in philosophy. That seems overwhelmingly the most plausible and, perhaps, the most fruitful attitude to take to the problem.

A second, quite distinct motive for probing one's theory with hopes for realism lies in a search not just for the restrictive but also for the *permissive core* of the theory. The restrictive core presents, in the fewest axioms, enough to generate all the theorems we think we want. The permissive core presents the axioms without which we could not reconstruct anything recognisable as *this* theory. Put like that, the cores may not sound obviously different and perhaps one might often reach the same axioms in each search. But they can, and in spacetime theory they do, diverge quite widely, motivated, as they are, so differently.

An example built out of material already at hand should make this clearer. It is very widely believed that a single principle about causality lies at the foundations of SR and GR. It is the Limit Principle: nothing outstrips light. We have been rehearsing the results of attempts to make this principle yield the whole metric congruence structure of spacetime in formal systems, attempts which make this the key element of their restrictive cores.

The nice things about the Limit Principle are that it is, very likely, true; that it is suitably qualitative so as to capture (by extension at least) the qualitative relation we have been looking for; and that it uses an ideology tied to agreeably familiar concepts like *particle* and *outstrips*. ('Outstrips' is preferred to 'goes faster than' since the latter is not transparently qualitative.) If we take this principle as a restrictive core (or a main part of it) we are likely to go on in our formalised spacetime theory to define the null cones at each point as the light cones there and, of course, to lay the principle down as an axiom of the formal theory.

We pay a price for this. It may bind theoretical imagination. Questions which might be fruitful speculations cannot arise. The hypothesis that there are tachyons, particles faster than light, is inconsistent with SR thus formalised and so is the hypothesis of the massive photon, since we have tied the conical structure of spacetime to the hypothesis that photons are massless. Yet both hypotheses, and more notably the former, are compatible with relativity (else they would never have got off the ground). Clearly, the restrictive core has its own modesty, hospitable though it may be to realism. It is no less diffident in denying structure than positivism is in affirming it. In the present case, and surely quite generally, permitting a structure (here the null cones) to belong to an entity epistemically remote (here spacetime) is precisely what allows us, within the theory, to speculate about the observable behaviour of epistemically proximate things. It opens the door to the design of new ranges of tests of the theory. What the permissive core permits is speculation *within the theory* as to tachyons and the like. The positivist may speculate about them too, but only by abandoning the theory which both disciplines and guides what he may conjecture and construct. Realism can be the springboard for flights of the empirical imagination.

When we ask for the permissive core, our question is quite differently motivated and our criteria for a good answer are not all the same as for the restrictive core. Though it is unlikely that either hypothesis just considered is true let us ask how SR would survive their truth. The answer is quite obvious once one glances at tachyon theory. The striking fact about this bit of physics is that the equations are all Lorentz invariant. That is they are the same for all inertial coordinate systems, since they remain unchanged, as all equations of physics do, under Lorentz transformations from one inertial frame to any other. So the permissive core of SR appears to be the Invariance Principle: that the laws of physics are invariants of the Lorentz transformations. What one has to say about the massive photon may involve some change in the laws of electromagnetism or optics. But what it most certainly does not demand is a change in the principle that whatever laws there are must be Lorentz invariant. That principle – the Invariance Principle – is the permissive core of SR. When the Invariance Principle is taken as the permissive core the constant 'c' in the Lorentz transformations cannot be interpreted as anything more than a finite inertial-frame-invariant scalar with the dimensions of speed. To interpret it as the velocity of light would preclude the massive photon hypothesis, and to interpret it as the limiting speed of causal propagation would preclude the tachyon hypothesis. Even to insist at this level that 'c' has its standard value of 2.9979×10^8 m/sec. would preclude further more precise measurement of the frame-invariant speed.

The distinction between the cores may become active in construct-ing formal theories (compare Sklar 1977a). We are indebted to this paper, but try here to take a different and in some ways, perhaps an incompatible perspective on the forces at work in constructing formal theories. In the present climate of speculation within physics, we are claiming that no one would seriously consider an equation which is not Lorentz invariant. To express it in the material mode, it may be active in our probing the ontic structure of the world. How? Well, SR is a very highly confirmed theory. It, or something very like it, must be true, locally at least. But *what* is highly confirmed, so virtually cer-tain that speculation inconsistent with it is really somewhat idle? Not the Limit Principle, certainly, strongly recommended though it plainly is by the positivist styles of critique as at the heart of the the-ory. What our willingness to countenance as speculation with SR – and surely the entertaining of alternatives has to count as a crucial part of creative scientific thought – what this marks out as basic is the Invariance Principle. Not only is this principle quite a different state-ment from the Limit Principle in being much weaker. It is expressed in an ideology which is richer in what is needed to state the principle itself. It asks for a richer ideology in the further axioms which must complement the Invariance Principle if we are to generate some interesting theorems.

It follows that there are at least two sorts of pressures in structural ontology – toward restrictive cores and limited ideologies and toward permissive cores and speculative ideologies. There appears to be no obvious principle which instructs us how to balance these opposing pressures (Sklar, 1981). But, of course, one wants neither the implo-sion of content so liable in positivism nor the explosion into pointless structures which results from reflex realism. We point out that a richer ideology does not automatically lead to our including bizarre theorems and a baroque ontology, either objectual or structural. However it does lead to our including theorems which may not differ from their contraries in observational consequences. It also liberates theoretical speculation, and we count this no less important than par-simony.

No mention has been made of the idea that theory structures are governed by choice of conventions. We find it unhelpful to divide the language of our theories into factual and merely conventional, non-factual sentences. This not only gives wrong answers to some questions but prevents the asking of others. In this context we can ask why we choose this convention rather than that to settle, say, simultaneity in a frame, and we can answer in terms of simplicity, convenience and so on. But we can hardly ask, for want of any idea how to answer, why we have a convention governing a unique choice of simultaneity in a frame but no convention singling out some one frame to be uniquely preferred though, plainly, such a convention is possible. If conventionalism were true surely the answers to such questions would be important and clear principles ought to govern them. But we see no room in conventionalism for such principles. If one choice is seen as an *assumption* about the world, we might at least hope to find a principle which explains how we choose, since assumptions, unlike conventions, are not necessarily arbitrary.

Further, there is nothing whatever arbitrary in the choice of space, time and spacetime, continuous and differentiable, as observationally undetermined arenas for the occurrence of physical things and processes. They are, if you like, virtually inescapable *a priori* concepts, though corrigible ones, which we bring to the organisation of our experience. That they are corrigible is abundantly clear in the physics of this century. The description '*a priori*' wears an antiquated look, to be sure, but not more than the ideas, to which it is no distant kin, of non-factual convention and of the observationally underdetermined as the observationally arbitrary. Just as it was the structure of the laws of classical mechanics that determined our assumptions about spatial congruence and not some observably invariant object, so it is the structure of laws in SR which determines the (local) null cone structure of spacetime and not some observable relation (signalling etc.). Just as an immense diversity of observations confirmed the former by confirming the laws, so it is with the latter.

(c) The Invariance Principle poses a new problem. There is something not quite right about this way of identifying the permissive cores. Something about the structure of spacetime is at the centre. For reasons already glanced at, the Invariance Principle obstructs a clear view of what it identifies because it speaks of extrinsic matters. The Principle is about frames of reference or coordinate systems and about transformations among these. We need to clear those cobwebs away. We can do so by saying something like this: spacetime has a null cone structure at each point within which there are no privileged directions. But reference to null cones imports quantitative ideology.

We want a qualitative principle and expect to find it in the permissive core.

In the light of all that, we suggest the following course. Given any point in spacetime, some set of points count as elsewhere (locally, at least). Also, there is a set of points which are *after* the given point and a set of points which it is *after*, as Robb noted. We do not need a relation which is asymmetrical, in the style of *after*. So let us speak of the *elsewhen* as what is before or after the given point. Then we have a quite elegant vocabulary of symmetrical qualitative relations which allows us to state, qualitatively, the conical structure of spacetime in pre-metrical language. The statement which lies at the permissive core of SR, then, says this:

> *The elsewhere and elsewhen of any point overlap in a (hyper-conical) boundary.*

Any spacetime point in this boundary, which is distinct from the origin point, is both elsewhere and elsewhen from it. There are no privileged directions in the elsewhere or the elsewhen. What we are claiming as the permissive core of SR is not merely something qualitative. It is also topological. The structure given in the statement is virtually the same as the structure of the ε-neighbourhoods in the fine (Zeeman) topology of Minkowski spacetime (Zeeman 1967). The ε-neighbourhood of a point consists of the manifold neighbourhood of the point, with the null cone through the point (though not the point itself) removed. That is to say, it is the union of the elsewhere and elsewhen neighbourhoods minus the boundary in which they overlap. Though these are not a *base* of neighbourhoods of x in the fine topology, they are undoubtedly important for it.

Thus the basic qualitative relation in spacetime is either elsewhere or elsewhen. That it is one rather than another of them has not been argued here. First, and most noticeably, the choice of either relation jars with the whole reductive trend in the positivist critique which aimed to make causality the basic qualitative relation, since each is, essentially, a geometrical relation. Causality is at least cashable in terms of matter, and it is moderately proximate to observation, but elsewhere and elsewhen are among those relations which have always given offence to reductive styles of thinking.

Which is more correct, or more searching, the positivist probe or the realist? That depends (among other things) on what you are willing to grant about what constitutes structural ontology. If structure is not an extensional study, then the positivist theory is incorrect as to questions about physical relations in the theory of spacetime. If structure is extensional, then positivism surely has much to be said for it. We argued earlier that ontic structure is not extensional, but surely those arguments touch the issue rather lightly and perhaps they hinge on special cases. They would repay more critical study. As for which probe is more searching, this depends on what you want to find. Both restrictive and permissive cores are worth finding, and if they are not the same for any theory, then each style of attack on the basis of a theory will reveal something important about structure.

(d) Finally, a rather brief glance at two further issues followed by a general conclusion. What Sklar (1977b) says 'might be right about the causal theory of time' is that the continuity or discontinuity of causal paths in spacetime are best taken to be basic, not in being the paths of possible causal connections but in being paths which might constitute the life-history of some consciousness. We have two doubts about this.

First, as Sklar himself mentions, it is not clear why this should give us grounds for preferring the elsewhen relation to elsewhere. We have a good basic intuitive grasp of local spatial continuity; we are, in fact, just as surely spatially extended as we are enduring in time. Further, it has often been noted that our temporal concepts tend to confuse us in thinking that time flows, that only the present is real and so on. This tendency, is due to the complex structure of temporal perceptions which is not easily described nor easily understood and is still imperfectly grasped (see §11, pp. 268–71). It is, therefore, less than suitable as a justification for our choosing continuous causal curves as basic.

Second, and more importantly, time-like curves are structurally basic in a way that space-like curves are not. In a two-dimension Minkowski spacetime there is a simple elsewhen construction dual to each elsewhere (or 'causal') construction, so that the Robbian programme can be carried out on the basis of elsewhere relations just as well. Less obviously, in four-dimensional Minkowski spacetime, where the dual to the timelike curve is the spacelike hypersurface of three dimensions, the programme is also possible on the basis of the relation '. . . is elsewhere from —'. (This claim is somewhat conjectural. We have not worked it through.) But there is an asymmetry, for while no two spacetime points are connectible by a time-like curve, in this spacetime, any two points are connectible by a space-like curve. In spacetime more generally, it is regarded as a pathology of the spacetime if this asymmetry is not preserved. The reason for the asymmetry is simple and basic. Time is one-dimensional and space three-dimensional. It is this structural, non-epistemological feature, which makes the time-like curve so simple and fruitful a means of probing spacetime structure. It is, in particular, a geometrical fact about spacetime and spatiotemporal relations, quite unsuitable for reductive purposes in being just the sort of thing reductionists want to reduce. So the basic qualitative relation that meets (or, perhaps, fails to meet) the formal requirements of Robb's programme is not a causal relation but a qualitative feature of spacetime itself.

If that argument is sufficient, then it illustrates how structural and objectual ontology may be connected. If the basic relation of relativity theory is elsewhen, or if the basic structure is the time-like curve, then a basic object in the ontology of SR and GR physics is spacetime itself. It must exist if the structure exists. Thus objectual ontology may depend, and we incline to the view that it standardly does depend, on structural ontology.[1]

1We are indebted in several places to discussion with Adrian Heathcote.

3 Special Relativity is not based on causality

1 Introduction

Almost everyone who has a considered opinion about what the conceptual foundations of SR (Special Relativity) might be, thinks that they lie in causality. (But see Lacey 1968, Smart 1969 and Earman 1972 for opposite opinions.) Cause is said to be the basic concept of the theory. The basic thesis (statement, postulate) of SR which makes use of this concept is said to be the *Limit Principle*: nothing (including causal processes) goes faster than light. This opinion about SR's foundations which I will call the *standard view*, is a philosophical one, not just because it is about conceptual foundations but also because many who hold it envisage the success of a programme attributed to Leibniz. It reduces the metaphysically uncomfortable ideas of space and time to the single familiar idea of physical material cause. This picture of the theory as specifying for itself how its bold new structure rests neatly on secure old foundations, as basic as can be, does much to shape our wider ideas about how people build their concepts and theories and about how these grow out of older concepts and theories.

Despite the widespread enthusiasm for it, and the careful and ingenious way it has been developed, the standard view remains doubtful.

The Limit Principle is very probably not even a consequence of SR and certainly not among its basic postulates. It is uncertain which concept causality is and whether the relation is fundamental to anything. I think that the standard view is false and that is what I will try to argue.

One sign that the standard view is dubious can be seen in the extensive literature on tachyons (a literature now in existence for twenty years) which has failed to show that SR forbids them. The idea of a tachyon is the idea of a particle which goes faster than light. The hypothesis that there are tachyons is flatly inconsistent with the Limit Principle. Thus, if the Limit Principle is a basic postulate of SR, there must be a brief, decisive refutation of the tachyon hypothesis simply by invoking that postulate. The literature on tachyons presents a very different picture to us. A rather strange, but consistent dynamics has been worked out for tachyons *within the context of SR*: mass, momentum and energy can be defined for them. Their oddities include that mass for tachyons is an imaginary quantity and that the temporal order of events in their histories varies under Lorentz transformations. Now, it is probable that there are no tachyons. It may even be the case that they can be shown physically impossible – because their quantum fields make no sense, say. But if the existence of tachyons is consistent with SR then the theory gives *no* particular significance to causality. Lastly, unless tachyons can be ruled directly out *by the Limit Principle* used *as a postulate*, then neither that Principle nor the causal relation has a *fundamental* significance in SR. What is intended by appeal to the status of being a postulate, being basic or being fundamental is looked at more closely in §3.4.

The crucial target which any relation proposed as basic to SR must hit is this: the relation must allow us to identify the light cone structure of spacetime. It is, of course, extremely obvious that the light cone is basic to spacetime structure and to SR. If the Limit Principle is a consequence of SR, then causality is coextensive with the surface and interior lobes of the light cone. More strictly, the relation expressed by '*x* is causally connectible with *y*' is coextensive with the relation expressed by '*x* is on the surface or in the interior lobe of the light cone at *y*' iff the Limit Principle is true. The Limit Principle identifies light as the limit of causality. If the Limit Principle is either false, or not a consequence of SR, or not a postulate of SR then, respectively, causality is not co-extensive with the light cone, not specified by SR as co-extensive, or not postulated in SR as the foundation of the light cone. In any of these cases the standard view would be mistaken.

Whether or not the Limit Principle has any status in SR it is certainly false that causality is the *only* relation which identifies the light cone or is said to do so, quite fundamentally, by SR. Clearly, but trivially, light (electromagnetism) itself does so. The relation '*x* is connectible with *y* by a (possibly reflected) light path' identifies the surface and interior of the cone. Rather more interestingly, a rather simpler and purely geometrical relation does so. It is 'occurs at another time than' (where this is understood to be determined without reference to any coordinate system or reference frame), the *else-when* relation. This is an agreeably direct and simple consequence of the theory in its earliest and most naive forms. There is, then, no obvious reason to think that SR will fail unless we use *causality* to identify the light cone. All that will collapse is the reductive, philosophical programme of defining spacetime relations by means of causal ones.

The Invariance Principle (that is, that the laws of physics, and of light propagation in particular, are invariant under transformations of the Lorentz group) is a far more prominent thesis of SR than the Limit Principle is. However, it is widely regarded in philosophical literature as laden with non-factual, merely conventional elements, and it is held that the Limit Principle exhausts its factual core. Both these views are false, I will argue. The Invariance Principle is not convention-laden and the Limit Principle forms no part of what it asserts. (By no means all who might describe themselves as causal theorists subscribe to this view. See Malament 1977b, for example.) Einstein seems to have been responsible for this widely held opinion. Something which is at least very like it appears in his 1905 paper. His discussion of simultaneity does seem to be based on the Limit Principle. His task here was to make the conceptual point that the apparently synthetic necessary truth that simultaneity is absolute could be undermined. He sees the Limit Principle as requiring that simultaneity in a frame be settled by a definition, a convention. What he goes on to say in embracing the Invariance Principle, however, is not that the finite speed of light is the limiting speed for every real process, but, in effect, that it is the *invariant* speed for all Lorentz frames. Evidently, he thought that the one thing entailed the other. It is not at all clear how it might do so. In the coordinate-free treatment of the metric structure of spacetime the crucial statement about the light cone is that it is null. Once again, this has no obvious consequence in relation to the Limit Principle. It is perhaps worth noting that the 'paradox' by which Einstein says he was led to SR actually suggests the idea of the invariance of the speed of light just as readily as it suggests it to be a limiting velocity. The 'paradox' runs thus: if one could equal the speed of a light wave, it would appear to one as a standing wave and no such thing is known. This suggests that the attempt to *approach* the standing electromagnetic wave by *approaching* the speed of light must also fail, so that light will appear the same for all attempts to approach it.

The consistency of the tachyon hypothesis with SR is of real importance for my argument in showing that the Limit Principle is not a consequence, certainly not a basic postulate, of SR. But it is not all important. I argue, also, that causality is too indefinite a concept to play the part which the standard view envisages for it. Attempts to explicate it run into various problems (especially the problem of circularity) if one wants to use it to reduce the ideas of space and time to causality. Again, to commit SR to causality as a basic concept is to make a problem of its consistency with quantum theory (though I shall not argue that at length). Lastly, I argue that we can define the light cone in ways quite independent, conceptually, of causality and even by a means of a relation which is not even coextensive with causality, whether the Limit Principle is true or false (or entailed by SR).

This makes my argument look as if it conflicts directly with what is probably the most interesting and penetrating work done in recent years on the conceptual structure of SR. Zeeman (1964), Malament (1977a), Winnie (1977) and others claim to define various structures in spacetime on a causal basis. For example, the Lorentz group (Zeeman), orthogonality (hence simultaneity) (Malament) and even the complete metric congruence structure of spacetime (Winnie) have been defined on the slender basis of causal topology. This work looks back to the early, but previously neglected, work of Robb (1914). The appearance of conflict is somewhat illusory, however. All these constructions can indeed be made on a topological basis, just as these writers say they can. But the basic relation of this topological structure is not (said by SR to be) causality. It has to be some other relation, either the elsewhen relation or the light-connectible relation itself, for example. But though I dissent from this fruitful body of theory on only a single issue, it is the fundamental issue.

In §3.2 I look at the role played by the Invariance Principle in certain standard formulations of SR, arguing that it carries much of the factual import of SR and that its supposed taint of convention is illusory and based on certain confusions. In §3.3 I glance at some simple, qualitative finds about tachyons and relate them to the Invariance Principle. §3.4 discusses various accounts of the causal relation arguing that some standard accounts would be unsuitable for their envisaged purpose even were the Limit Principle a consequence of SR. Lastly, in §3.5, I consider whether the surface and interior of the light cone has any deeper significance for spacetime structure than the surface and exterior has. I conclude that it has not, which finally demonstrates that causality can have no *unique* significance for SR, whatever the status of the Limit Principle might be.

2 *The Invariance Principle*

A rather satisfactory picture of the import of SR is given by the claim that the Invariance Principle is a central postulate of it and that this Principle does not entail the Limit Principle. An advantage of this picture is that it is cast in rather familiar ideas which make it clear how and where it conflicts with the standard view. Since that part of the physics of tachyons which is of most interest for the matter in hand is drawn out of the Invariance Principle, and this physics shows that the principle is independent of the Limit Principle, it will be useful to look with some care at what Invariance is. But though this way of viewing SR is rather satisfactory I will say, at the end of the section, why I do not believe that it offers us a suitably *basic* picture of the factual structure of SR.

What is the Invariance Principle, more explicitly? How clear is it that it is not a partly conventional, non-factual, statement? To answer either question we must begin by distinguishing two styles of interpretation of SR and see that the Invariance Principle plays somewhat different conceptual roles in each. By and large, it does no harm to run the two styles of interpretation together but this is so only if one does not think that the Invariance Principle is some sort of convention. What are these two styles of interpretation?

I call the first the *spacetime interpretation*. It has an ontology of events (alternatively of four-dimensional objects) and speaks of a spacetime (but not of a space nor a time). Spacetime in SR has zero curvature and Minkowski metric in which physical quantities appear as 4-vectors or 4-tensors. A very wide range of coordinate systems may be used in the description of spacetime and its physics, but assigning the same time coordinate (or the same space coordinates) to different events is not interpreted as reflecting, even relatively, the concepts of simultaneity (or rest) since these are no part of the scheme of concepts of this interpretation. Further, the use of coordinate systems is not essential to spacetime physics, coordinate-free treatments of it being quite standard. Light is distinguished neither as invariant in speed nor as the fastest signal, since ideas like 'speed' and 'fastest' do not belong in this interpretation. Light is picked out in that light trajectories always lie in cones of null geodesics, one of which is defined at each spacetime point.

The second style I call the *relativity interpretation*. It has a familiar, classical ontology of continuant spatial objects persisting in time. It uses the idea of a frame of reference, a relative space and a relative time. The frame defines (relatively) the *same* place at different times and thus the concepts of rest, motion, speed and velocity. It also defines (relatively) the same time (simultaneity) at different places. Light is marked out as having an *invariant speed* for every frame of the interpretation. Whereas one can do spacetime physics without coordinates, it is meaningless to speak of a frame-free relativity physics. The concepts of the relativity interpretation make sense, in SR, only with respect to some frame. For these, and other, reasons the idea of a frame of reference does not correspond in a simple way to the idea of a coordinate system. Consequently the part played by the Invariance Principle is a little different in each interpretation. What are these differences?

It is widely agreed that the spacetime interpretation of SR is the more fundamental one. Two principles about coordinate systems for spacetime make sense in this interpretation. The Principle of General Covariance says that any coordinate system for spacetime is admissible so long as it is continuous and distinguishes time-like from space-like coordinate trajectories. So it admits skew (*time*-skew) and curvilinear coordinates as well as the familiar Lorentz linear, time-orthogonal coordinates. The Principle says that the laws of physics are covariant (or invariant) under the very general group of coordinate transformations which map from any one of these systems to any other. This really amounts to treating spacetime and the vectors and tensors in it as geometric objects which are what they are irrespective of the coordinates – mere devices of book-keeping – which are used to describe them. The form of any law which is invariant under this general group of transformations is not very simple. It contains terms which are functions of the affine connection and metric tensor of spacetime. The Invariance Principle is a more special statement about a particular set of coordinates within the general set and about the particular group of coordinate transformations which connect them. The particular set is, of course, the linear, time-orthogonal (Lorentz) coordinates and the particular group is the Lorentz group of transformations. What the Invariance Principle says is that the form of laws invariant under transformations of the Lorentz group is really very simple, for it need contain no terms referring to the affine connection or metric tensor. That constrains the laws of physics in a rather significant way. In flat spacetime, at least, General Covariance places no additional constraints of a physical kind on the laws. (See Weinberg 1972, pp. 92–3 for a clear discussion of this.) Obviously, the fact that the set of Lorentz coordinates and its group of transformations are included in the more general set of coordinates and its groups of transformations does nothing to *weaken* the constraints imposed by the Invariance Principle on the laws of physics. Thus it seems apt to focus on this principle as elegantly conveying a main factual assumption of SR. Of course it figures very prominently in the structure of SR.

Clearly, in the spacetime interpretation, the Invariance Principle does nothing to restrict the range of admissible coordinates. It does not forbid the use of skew or curvilinear coordinates. There is no sort of conflict between it and the Principle of General Covariance.

In the relativity interpretation, by contrast, the Invariance Principle does function as restricting the choice of reference frames. It says that only Lorentz (or inertial) frames may be used. Where does the Principle of General Covariance fit into this interpretation, then? The relativity interpretation does not let us formulate a principle which might be properly seen as a counterpart of General Covariance. Partly, this is because the idea of a reference frame is really quite different from that of a coordinate system. Frames differ just when they define different *spaces* (sets of *rest* points) or times (set of simultaneous events). So the ideas of a space, a time, of rest and simultaneity go inextricably together with that of a frame. However, a mere shift of origin, or a purely spatial rotation of space coordinates results in a new coordinate system. So frames correspond, at best, to *classes* of coordinate systems. However, beyond all this, the ontology connected with the idea of frames is different from spacetime ontology. This means, for present purposes, that whereas terms involving the affine connection have a purely geometric sense in unified spacetime (which is what permits General Covariance), the split into a space and a time needed for frame of reference ideas prevents these terms from carrying over a merely geometric meaning. Though one gets the same laws formally in an accelerated frame as one gets in the corresponding curvilinear coordinates, by involving the metric tensor and the affine connection, one has to fabricate inertial forces and introduce dynamical concepts, to deal with these terms. These make no more physical sense in SR than they did when classical physics considered the consequences of using accelerated frames. For these reasons, the Principle of General Covariance is conceptually prevented from having any proper counterpart in the relativity interpretation. So we are obliged in that interpretation to read the Invariance Principle as restricting the class of admissible frames.

In this way, the two interpretations give somewhat different pictures of where Special theory ends and General theory begins. The relativity interpretation makes it look as if General Relativity starts with the extension to 'accelerated' frames. But it is not really here that the important generalisation comes. As the spacetime interpretation shows, it is in the extension to curved spacetime with the entry of gravitation that General Relativity begins.

These observations on the Invariance Principle would not be of much importance were it not widely held that this Principle is really a convention, with the Limit Principle as its factual core. The view that Lorentz simultaneity is a convention within a frame of reference is often tied closely to the idea that the Limit Principle is basic to SR and that causality is its fundamental concept. In the light of that view, it seems to some that the tachyon hypothesis must completely subvert the relativity of simultaneity, since if the Limit Principle did not obstruct attempts to set up absolute simultaneity, nothing else could do so. But this overlooks the role played by the Invariance Principle in tachyon physics.

Of course, one can hardly see what the role of the Invariance Principle is if one thinks that it has nothing factual to say. Conventionalists in respect of simultaneity think just this. I argue for the view (§3.4) that simultaneity in a frame of reference is no matter of convention and that the view that it is rests on some confusions.

To sum up this section so far: the Invariance Principle is a main factual assertion of SR, best understood in the spacetime interpretation as placing stringent constraints on the form of physical laws under a central group of coordinate transformations. The Principle is innocent of conventional taint (in the sense of containing non-factual elements). The theory that it requires a convention of simultaneity springs from confusions about the relation between coordinate systems in the spacetime interpretation and frames of re _ ₌nce in the relativity interpretation. This explains and defends the positivist view of the foundations of SR which I state in the first sentence of the first paragraph of the section.

Nevertheless, there are reasons for thinking that the Invariance Principle is not quite the right replacement for the Limit Principle. First, the Invariance Principle tells us something crucial about certain *coordinate formulations* of physics for spacetime. But facts about its coordinate systems can hardly be what is basic to spacetime and, indeed, one can do spacetime physics without the use of coordinates at all. Second, it would be nice not to have to concern oneself with the rather inelegant relationship between General Covariance and Lorentz Invariance, even if the first point were waived. We can avoid the relativity to coordinate systems or reference frames (together with some of the problems just discussed, though some illumination is gained by discussing them, I hope) by citing the metric facts about spacetime; that light geodesics are null, space-like intervals (Δs^2) negative and time-like ones positive (for appropriate signature) and, generally, by describing the metric tensor of spacetime. It is, in fact, the metric structure of spacetime which bestows on the Lorentz coordinates (and their connecting group) the special symmetries which make possible the simple forms of physical law. It makes the times and spaces of the frames of reference corresponding to these coordinates simple and isotropic. Last, the Invariance Principle is framed in the concepts of the spacetime *metric*. Now the Robbian constructions of Zeeman, Malament, Winnie etc. mentioned in §3.1 strongly suggest that metric concepts are not basic to SR (but see Sklar 1977b).

One only needs the topological structure of the conical ordering of spacetime. Then one can *derive* the whole metric congruence structure.

I propose, instead, that a fundamental relation of SR is the elsewhen or non-simultaneity relation ('*x* occurs at a time other than *y*'). The basic assertion which uses the relation is that *x* occurs at a time other than *y* iff *(x ≠ y)* and *x* is in, or on the surface of, the light cone at *y*. This is suitably topological. There is no other relation which, *according to SR theory*, allows us to identify this relation and thus reduce it (save for light connectibility itself). I will argue later, however, that SR can be based in a different way on a quite distinct relation, with equal satisfactoriness.

In the next two sections I will frame the discussion in terms of the Invariance Principle, since I think this is more familiar and easy to follow.

3 Tachyons

This section considers the physics of tachyons rather briefly and roughly. In the course of it, I argue for these conclusions: it would be reasonable to expect that tachyons, if there were any, would be useless for signalling or synchronising clocks; nevertheless it is not inconsistent with the Invariance Principle that they should have such a use; their use in synchronising clocks could not override the relativity of simultaneity; these facts about signalling do not disqualify tachyons as causal particles; at least one attempt to disqualify tachyons as causal particles appears also to attempt to deduce the Limit Principle from the Invariance Principle, contrary to causal theorists' views that the former principle is, relatively, the more basic.

A tachyon is a particle which goes faster than light. If a particle ever does so, then it always does so. A glance at the relativist formula for kinetic energy makes it clear that the energy of a particle increases without limit as its speed v approaches c (from $v < c$). One could never accelerate a slower than light particle to superluminal speeds. If $v > c$, then both the energy and momentum become imaginary quantities, if the mass remains real. But if an imaginary (complex) quantity is admitted as the mass of a superluminal particle then the energy and momentum remain real though they increase without limit as v

approaches c (from $v > c$). Thus a tachyon is always superluminal just as ordinary particles are always subluminal. This entails that there can be no procedure of measuring the mass of the tachyon at rest and we may suppose it to be imaginary. Thus the velocity and energy-momentum 4-vectors of these particles are always spacelike at each point.

The Invariance Principle tells us not only which properties of tachyons are frame-invariant but also which ones are not. This negative aspect of the message is more important for the argument of this section. For example, the velocity and the energy-momentum 4-vectors for tachyons are space-like at every point of a tachyon worldline. It would seem, therefore, that the energy (that is, the time component of the latter vector) will be negative in some frames, though positive in others; further the velocity vector may have a negative time component, which appears to mean that 'it travels backwards in time'. However, if we reflect more carefully on what the Invariance Principle implies, we see that it requires that events in the history of a tachyon do *not* have an invariant time order: that relative to some frame A, the tachyon is *emitted* at E_1 and travels to the point of its (frame-relative) later *absorption* at E_2, whereas relative to another frame B it is *emitted* at E_2 and travels to the point of its (frame-relative) later *absorption* at E_1. In the light of this reinterpretation of the (relative) temporal ordering we may reinterpret the energy of the particle as positive in each frame also. But, then, the tachyon is described in frame A as an antiparticle of that described in frame B. This device is called in the literature the Reinterpretation Principle. It shows how the Invariance Principle constrains the physics of tachyons so that they become consistent with SR.

The laws of tachyon physics must be Lorentz invariant if they are to be compatible with SR (and only the possibility of a physics consistent with SR concerns us at the moment). Thus no law or laws of tachyon physics can specify conditions sufficient for the emission of a tachyon by a subluminal particle since, if such an event occurs in any frame, there will be infinitely many others in which the same conditions suffice for the *absorption* of a tachyon. So, first, there can be sufficient conditions only for the emission-or-absorption of tachyons by subluminal particles. It is best to speak, in spacetime language, of laws specifying conditions for the existence of a tachyon *pencil* (or one or more *rays*) at a certain spacetime point on a subluminal particle, this being neutral as to which are emitted and which absorbed. Secondly, though this does allow for the interaction of super- with sub-luminal matter (without which the hypothesis of tachyons seems empty), it strongly suggests that the conditions sufficient at t_1 for a tachyon pencil at $t_1 + e$ on particle a cannot be exhausted in conditions *local to a*. How can I tinker with a at t_1 to make it absorb a tachyon at $t_1 + e$ unless there already is a tachyon elsewhere at t_1 free to be absorbed?

This last point is not at all straightforward. Consider the case of the light cone. Nothing in SR entails that light is characteristically emitted in expanding spherical waves rather than characteristically absorbed in contracting ones. Nevertheless, that is invariably the case. This asymmetry between absorption and emission (the preference for retarded over advanced solutions of wave equations) seems to be related, at a deep level, to the directedness of time. It has no explanation in terms of known fundamental laws, all of which are time-symmetric and I think it fair to say that it is still not well understood. Popper's attempt to explain it by appeal to improbability is circular. He argues that the contracting wave front culminating in the event of absorption requires an immense, non-local conspiracy of conditions at each point (at some time) on a sphere with the absorption point as centre. This conspiracy directs radiation inwards to this central point. But why is this a conspiracy of events of an immensely improbable kind, whereas the time-reflected existence of a spherical radiated wave front moving outwards from a point of emission is not? Simply because we always get the retarded solution, never the advanced one. If we did always get the advanced solution, as the laws allow perfectly well, we would *not* regard it as *at all* coincidental. A similar objection holds if we understand Popper's appeal as directed toward the simplicity of the point of emission as a proper locus of explanation. If we always got the advanced solution then we would come to regard the subsequent *absorption* of the spherical wavefront at a point as explaining, in a simple way, why there was a collapsing spherical wavefront preceding it. We might now argue for the conspiracy of events by appeal to what is, essentially, our sense of the direction of time. This appeal would be somewhat inarticulate, since it is unclear whether this raises anything independent of what we want to explain or whether the appeal refers to anything well understood. The analogy with the tachyon example is imperfect, but it suggests that assuming localised control of tachyon interactions, which is something essential to signalling, surely, is not at all straightforward and needs to be considered very cautiously.

If there can be no law-like invariant distinction between the emission and the absorption of tachyons, then it would seem that they cannot be used in any direct way to send signals or transmit information. This conforms to what one might expect. But it is not, I think, a direct entailment of the Invariance Principle. The ideas of a signal (a mark, information) and the sending of it are, for physics anyway, ideas of irreversible processes. They belong in thermodynamics. One might expect that the 'laws' of thermodynamics, like the more fundamental time-symmetric laws of SR proper, will be invariant under Lorentz transformations. If that is so, then irreversible processes will be irreversible in the further sense that their temporal order will be invariant under the Invariance Principle. In this 'natural' sense, then, it will be the Invariance Principle (not the Limit Principle) which confines signalling and information transfer to the light cone. For this reason, too, the causal paradoxes sometimes associated with tachyons (e.g. Salmon 1975, pp. 122–3) will not arise, since they presuppose that one can use a tachyon to signal. (See Feinberg 1967, Appendices A and B, pp. 1101–3 for a more detailed account of the emission – absorption and signalling problems.) Again, for the same reason, tachyons can have no simple use as devices for synchronising clocks.

Of course, the relevant 'laws' of thermodynamics might not be constrained by the Invariance Principle, being statistical, possibly local in spacetime and in the nature of boundary conditions rather than true laws. If so, things are not so simple. There might be, for each spacetime point, a sort of thermodynamic (space-like) tidemark in spacetime so that tachyons could be marked only if their velocity vectors point, towards the forward (dry) region of spacetime beyond the tidemark at that point, and marks erased from them only if their velocity vectors point towards the backward (wet) region of spacetime. (The tidemark metaphor is not intended to suggest that a tide moves in spacetime.) Then, it might seem, tachyons could be used to set up a *de facto*, statistical sort of simultaneity, but an absolute one nevertheless.

But this would leave SR and all its conceptual structure quite untouched. It would still be true that the fundamental laws of physics were Lorentz invariant. It would still be true that everything observed in every Lorentz frame (that is, *using standard light signal synchrony*) would be physically possible and, for much the most part, perfectly as usual. The statistical, *de facto* tachyon simultaneity would, perhaps, mark out some frame uniquely as the one in which this simultaneity coincided with standard signal synchrony. But nothing connected with true physical laws and their invariance would oblige us to prefer this frame and it would, probably, not even be strongly recommended by convenience (unless we could, somehow, *see* by means of tachyons). Statistically irreversible processes would be irreversible in only a weak sense, since their reversals would be observed in an infinite class of frames of reference which our fundamental theory of space and time assures us are admissible.

If tachyons are to be cast in the role of causal particles so that tachyon connections are causal connections, we need to countenance a symmetric relation of causal connection. In fact, causal theorists regularly do so. This will be looked at a little more carefully in the next section. But, quite apart from this, the case for regarding tachyon interactions as causal looks very strong. Tachyons have mass; they have energy and, in their interactions with bradyons (subluminal particles), they exchange energy. A tachyon interaction may change the momentum of a bradyon either by accelerating it or by being absorbed by it. This is, surely, unequivocally causal.

It is not my place to give more than this rather casual glance at the tachyon literature. More extensive discussions are cited in the bibliography.

4 Causality

Some obscurity surrounds which relation is intended by references in the literature to causality and which features make it proper to describe a relation as causal. More modestly, the literature leaves me uncertain how to answer some pertinent questions about cause; thus it leaves me unsure just how the causal theory can fairly be criticised in some of its aspects. How is the idea of causality, when it is proposed as a basis for SR, related to the idea examined by Hume and found so deeply problematic in philosophy ever since? When are we to understand causality itself as being *explicated* or *analysed* in the literature rather than merely indicated or identified, and how may accounts of it run the risk of making circular the attempt to reduce temporal relations to causal ones? In what way is causality proposed as primitive? In what way does it define, explicate or give the physical basis for temporal relations?

One problem which seems clearly urgent for the causal theorist is to fix *non-local* relations, to connect events here and now with events there and then, without calling on temporal relations to do so. So what seems to be needed in the idea of cause is only the part of it concerning causal *propagation*, to be understood somehow independently of temporal ideas. To focus just on causal propagation has two advantages. First, it lets us avoid the thorny question how to distinguish cause and effect, a question which arises for local as well as non-local causal relations. So causal *connection* is the key idea, and it looks agreeably free of Humean complexities. Second, this relation can be understood as a symmetrical one, raising no questions about the direction of time or cause. This seems appropriate to the time-symmetric structure of SR basic laws and to its (widely agreed upon) lack of comment on temporal direction. It is unsuitable to look for a basis for SR in an asymmetric relation of cause. (See Mehlberg 1935 and Grünbaum 1973 for critiques of Reichenbach's 1958 use of the asymmetric mark method also discussed in (*c*) below.)

There are deeper reasons why we should allow a fundamental relation of causality which is symmetric. We want to leave room for theories of the *direction* of time (and cause) which base it on contingent facts – not just logically contingent ones, but facts that are contingent in the light of fundamental physical laws. A theory of this kind is the Boltzmann–Reichenbach–Grünbaum theory, in which the direction of time depends on the feature that the entropy of almost all thermodynamic branch systems runs from greater to less in the same temporal sense. This is thought to lie in something like a boundary condition, which may apply only in some (large) region of spacetime and which is no matter of physical law. It allows that, in other regions of spacetime, the condition may fail either because the vast majority of branch systems have their entropy running from greater to less in the opposite temporal sense to that in our own region, or because the branch systems do not, on the whole, increase their entropy in a common temporal sense. Such epochs may coincide with periods of expansion, contraction or stability in the size of a spatially finite universe, for instance. In any of these epochs mechanical interactions would comply with those same laws of continuity, contiguity, motion and energy exchange which make up the core of the idea of cause. So it seems right to speak of a basic idea of causal connection which is symmetric, not just in the weak sense that it is given by a disjunction (*a* causes *b* or *b* causes *a*) of sentences using a deeper asymmetric causal relation, but in the strong sense that it captures what is given in the time-symmetric laws fundamental to physics. It is in this symmetric sense that tachyon connections are to be understood as causal. That there is this symmetric idea of cause, and that it is basic is, I think, perfectly acceptable to causal theorists and, indeed, defended by them.

This feature of symmetry at least partly explains, I think, why explications or analyses appear to be offered for causality, though it is proposed as primitive. The relevant aspect of it has to be distinguished from cause more generally. More importantly for its aim in giving a basis for temporal relations, we need to see how to specify the causal relation, *at least in extension*, without recourse to temporal (or to spacetime) relations.

There are at least three ways in which a property might figure as a primitive basis for a theory. It might be a *conceptual* primitive, an idea so clear, well entrenched and unproblematic that no account of it will make it more readily understood. Next, though it might be imperfectly understood, it might nevertheless be so basic to the structure of the theory in question that no formulation of it could both give us the theorems which identify the theory yet fail to give the relation, and key assertions using it, a very prominent place. Thus one could not understand the theory's other main ideas without understanding the first one at least equally well. We might call such a property a *theoretical* primitive. Lastly, it might be an *ontological* primitive; that is, a property so basic in the structure of the world (but not, necessarily, of the theory) that, despite our, perhaps dim, insight into it, there can be no more certain way of showing where the theory touches reality, so to speak, than to tie its ideas to this ontological primitive. It is clear, I think, that the standard view is committed to causality as a theoretical primitive of SR. It is not always clear to me how far the other (independent but compatible) notions of primitiveness are at issue in the literature. Whether one sees temporal relations as analysed or theoretically explicated by causal relations or the latter as providing the physical basis for the former may depend on the kind of primitiveness causality is reckoned to have.

Let us now look at several of these accounts of causality.

(a) In Reichenbach's (1958) and (1971), Grünbaum's (1973) and (1968) and van Fraassen's (1970) causal theories of time in SR, the notion of *genidentity* plays a very prominent role. This is not the same idea as causal connection, but is proposed as one kind of it. In Grünbaum's constructions, at least, it seems to play a more prominent role than causal connection itself. One can indicate what genidentity is in the following way:

> e_1 is genidentical with e_2 iff (i) $e_1 = e_2$ or (ii) ($e_1 \neq e_2$) and e_1 and e_2 happen to (are events in the history of) one and the same material thing.[1]

(That this takes the form of a definition is not intended to insinuate that the idea is not primitive in any of the senses just described. The first disjunctive clause in the definiens could be dropped without real loss.) We can go on to indicate the intended extension of the relation of causal connection, as van Fraassen does (1970), p. 179, by saying that two events are causally connected if they are genidentical or connected by a light ray.

Why does genidentity figure so prominently in these theories? It is not perfectly obvious that genidentical connections really are *causal* ones. How clear is it that the existence of something a while ago is the cause of its existence now? But I doubt that this query matters much. It seems to me that the appeal of this idea is that it enables us to speak of non-local connections without speaking of propagation as such; that is, without reliance on spatiotemporal criteria. Genidentity also seems to be used to display features, such as continuity, which causal connections (chains) generally show. In support of this interpretation of the role of genidentity, one finds van Fraassen saying that the 'single basic strategy' of Reichenbach and Grünbaum 'consists in *first* explicating the time order of the events on a single world line and then explicating the time order of all events by correlating the world lines (through the relation of causal connectibility)' (1970, p. 182, italics in original. His own procedure is to reverse the order of this strategy). In fact, as he says on this same page, the latter task is more easily understood as carried out by using the relation of topological simultaneity (causal non-connectibility). Thus non-local con-

1 Cf. Grünbaum (1973), p. 189; (1968), p. 57, van Fraassen (1970), p. 34.

nections are set up by identifying world lines but not using the concepts of time or time-likeness to do so. The Limit Principle then says that all worldlines lie within the light cone.

Genidentity has pleasing features as a primitive. It connects neatly with the very familiar idea of a continuant thing. It is clearly symmetrical. It makes no appeal to a prior concept of propagation or of time. It looks well qualified for conceptual primitiveness.

However, there are decisive objections to genidentity as a primitive, in the present context of use, at least. First tachyons count as material things. So unless SR says that there can be no tachyons, it does not say that genidentity (hence causality) is confined to the light cone. So genidentity is not theoretically primitive and SR makes no fundamental assertion about it which would allow us to use it to identify the light cone.

But there are more damaging objections than this. Genidentity must identify exclusively time-like relations if it is to enable us to pick out worldlines. It does not. Distinct events which happen to a voluminous material object, such as the earth, may well be simultaneous, hence space-like connected. Of course, the earth is really an aggregate of more fundamental material things, and genidentity may be intended here to apply only to these. But still, unless the idea of a fundamental material thing entails that every such thing is a strict point particle, genidentity cannot distinguish time-like connections from space-like ones. For distinct events which happen to it could be simultaneous (in some frame) and occur in different places (for every frame). Now, first, it does not seem to be true that fundamental material things are actually strict point particles, nor that any coherent notion of genidentity or causality can be ascribed to them (see Reichenbach 1971, section 26, and Smart 1969, p. 391). So genidentity does not appear to be an ontological primitive. Quite apart from these quantum problems, it is certainly not the case that SR anywhere claims or presupposes that some kind of atomic, let alone point-particle, hypothesis is true. It can be, and standardly is, expounded in terms of macro, spatially extended things (rods and clocks) and light treated as an expanding spherical wavefront. It is, principally, a physics of fields, and it would fundamentally misrepresent the theory to tie it, conceptually, to a strictly particulate hypothesis about the structure of matter. Thus, on this ground too, genidentity *in the sense required* does not seem to be a theoretical primitive of SR. A last weakness in this understanding of genidentity is that it makes a silent appeal to a spatial concept. Unless fundamental things are conceived as spatial points, as continuants with *only* temporal extension, genidentity cannot identify the timelike relations it is required to pick out. Thus, for the Leibnizian programme of basing spacetime (spatial as well as temporal) relations on causality, it is not a conceptual primitive. Both Grünbaum (1968) and van Fraassen (1970) gave up the programme for spatial relations though Winnie's (1977) constructions must surely revive the hope of carrying it out. But genidentity cannot be used without circularity. Thus this first version of causality is in deep trouble quite apart from the problems of the tachyon hypothesis.

Grünbaum (1973, p. 189) says that genidentity is, intuitively, 'the kind of sameness that arises from the persistence of an object for a period of time'. (See also his 1968, p. 57.) This understanding of genidentity might be used to free it from any appeal to the idea of a point particle concealed in the version just discussed (though I doubt whether this was Grünbaum's reason for giving the quoted comment). But this does clearly lean, quite fatally for *conceptually* reductive purposes, on the unanalysed idea of persistence in time. That cannot be a *definition* in the course of a Leibnizian reduction of the concept of time.

(b) By far the most common, brief statement of what causality is, says that events are causally connectible if a *signal* could be sent from one of them to the other. The Limit Principle then says that light is a first (fastest) *signal.* It is doubtful whether anyone thinks that this adequately explains the sort of primitiveness causality has if it is to play the role for SR envisaged in the standard view. But let us see, first, what advantages this account might seem to offer.

One of the features of tachyons, as we saw, is that they cannot be used to pass information or to signal (on the natural assumption that the 'laws' of thermodynamics are Lorentz invariant). This 'first signal' version of the Limit Principle would seem to exclude tachyon-connected events as causal since the signalling relation would coincide with the forward light cone. This suggests, at first glance anyhow, the idea that, since *information* (needed for setting clocks, for instance) cannot be sent from here to there any faster than light, we can never know how to synchronise clocks here with clocks there. Thus the relativity of simultaneity would rest on an *epistemological* fact.

The weakness of suggestions like this is, as before, that they really do not seem to be basic to SR in the way wanted. The idea of a signal is really the idea of a kind of irreversible process which transmits information or a mark. Thus it is an idea which belongs to thermodynamics and not at all, fundamentally, to SR. All the basic laws of SR are time-symmetric, unlike those of thermodynamics, and are incapable of specifying irreversible processes and, therefore, of capturing the idea of a signal. This puts the matter rather simplistically, but the point seems very widely accepted. I include it here for the sake of completeness rather than as controversy.

(c) In other places, genidentity is treated as if it were not a primitive relation on which causal connectibility is based. Instead, both these relations are defined in terms of others. Reichenbach (1971, pp. 224–7) appears to *define* material genidentity[2] as the prelude to his arguing that the relation does not apply to particles in quantum mechanics. He bases material genidentity on three criteria: that change is continuous; that any one particle is spatially excluded by any other; that the interchange of spatial position is verifiable (by marking objects). In reply to an earlier suggestion of mine that causality might not be fundamental to SR, Salmon (1982) refers to various discussions (1975; 1977a) he has given as to the nature of causal processes and what distinguishes them from pseudo-processes. One finds there that criteria are offered for causality which are analogous to (improvements on) Reichenbach's. Briefly, Salmon proposes both spatiotemporal continuity and the ability to carry a mark as criteria for causal processes. It is the latter which distinguishes causal processes from pseudo ones. As I understand him, Salmon wants to develop this latter idea in such a way that we can understand causal processes in a time-symmetric way, and thus he avoids the problems just discussed in (b) above. This makes for a very interesting development of the idea of cause.

Before pursuing it in more detail, however, I will point out, briefly, two difficulties which any account along these lines must run into if it wishes to show how causality in SR allows us to realise Leibniz's programme. First, Reichenbach and Salmon make it a necessary condition for a process to be causal that it be spatiotemporally continuous. But this idea already requires *at least* that we can understand and

2 This, rather than functional genidentity, is what concerns us at present.

identify what it is for two events to occur at different times or at different places from each other. A concept of causality for which spatiotemporal continuity is a necessary condition would thus be disbarred as circular from serving Leibniz's aim to reduce spatial and temporal relations to causal ones, whether this is intended to identify the relation intensionally or extensionally. Second, material genidentity and causal process break down for quantum phenomena as Reichenbach (1971, pp. 224–7) makes quite clear. But, if that is so, causality does not really seem to be a fundamental relation in the world at all, and so no fit candidate as the ontological primitive basic to SR. Further, to view causality as basic to SR is to view the theory as inconsistent with quantum mechanics in a basic way. No doubt the problems of quantised gravitation in General Relativity must be recognised. But there does not seem to be real conflict between SR and quantum theory. It seems an unfortunate ambition to create one by insisting that SR be based on causality.

Salmon's main idea of causality, however, is the ability to transmit a mark. He makes it clear that he is not relying on a prior understanding of irreversible process, of imposing as against erasing a mark. Thus he is proposing something quite distinct from the idea that causality is the asymmetric relation of sending a signal. It is the capacity to retain (and hence transmit) a mark which Salmon is emphasising.

I do not want to try to debate the main point that Salmon was making in these papers: that is, a basic ingredient in the style of scientific explanation which is familiar to us is the idea of a causal process, and the existence of such processes is a contingent fact (if a fact at all) about the world. Whether that point is sound or not is irrelevant to what I am arguing.

Salmon does not consider at length any possible objections to the idea that causality in this sense is coextensive with the light cone. However, he certainly claims that it is, in several places (1975, p. 141; 1977a, p. 217). This claim seems adequately answered by the brief discussion in §3.2 of the present paper and in the tachyon literature more widely.

But these remarks do not directly address the most interesting aspect of the suggestion which Salmon proposed. The suggestion does not really depend on ideas like signalling, or engraving as distinct from erasing a mark, and so on. Quite clearly the core idea is that the mark will occur at all points in the causal process between A and B. It does not matter, for the purposes of the criterion, whether it is being transmitted from A to B or from B to A. The question we have to scan is whether SR tells us that this criterion can somehow be used to identify the light cone.

If I have understood Salmon correctly, then it should make no difference to the criterion whether the feature which occurs at all points between A and B without further local intervention is a reversible or an irreversible feature – spin in the one case or a scratch in the other, for example. I will make that assumption. I assume, too, that we can discuss the proposal adequately in terms of classical concepts and ones which apply to macro material things. On the one hand the quantum complication can hardly aid Salmon's causal argument, and on the other it is not clear how to mark either a quantised or strict point particle. So let us look on the tachyon as a little markable billiard ball.

The following ideas then make perfectly good sense within SR. A tachyon is involved in a local interaction at a spacetime point A and a local interaction at a distinct point B. There is no other local interaction between the points. At all points between A and B the particle is spinning. With respect to some frame F, the interaction at A causes the particle to spin until this is braked by the interaction at B. With respect to some other frame G, the B interaction is the cause of the spin, which is braked at A. (The basic idea of spin as a reversible process with 'reversible cause' comes from Reichenbach 1971, pp. 184–6, of course). I claim, as before, that SR asserts nothing inconsistent with the capacity of tachyons to transmit such a mark, consequently it says nothing which would allow one to exploit this capacity to carry a mark as a criterion which could be used to identify the light cone.

(d) It seems appropriate, next, to look at the pioneering and, by now, influential work of Robb (1914). While Robb's influence has been suitably acknowledged, there seems to be little discussion of the materials which he actually worked with.

Robb's primitive relation is *after*. He defines `before' by means of it, so that it will do the same work as the *elsewhen* relation which I used at the end of §3.2. He postulates that *after* is asymmetric; that there is at least one element after any element and so on. However, in the Introduction to his book Robb offers some remarks about this primitive as if he felt obliged at least to tell us which relation it is. Consequently, there is some uncertainty as to whether *after* really is the primitive of Robb's system or whether the ideas used in his introductory remarks should not be preferred. This uncertainty appears to be the only basis for taking Robb as a *causal* theorist.

A glance at the ideas Robb makes use of in his Introduction and which he sees as shedding some light on the idea of *after* strongly suggest philosophical motivation. A key remark is the following (p. 4, his italics):

> Of any two elements of time of which I am *directly* conscious one is *after* the other.

(I guess that Robb also meant to imply that one is aware of the one event *as* after the other.) He goes on to claim that the set of instants of which any one subject is directly aware form a series in linear order (the ordering relation being, I guess, '*S* is aware of *x* as after *y*'). Evidently, Robb wants to tie the *after* relation to what he sees as a solid epistemic anchor. He later extends the relation by using the following condition (p. 7):

> If an instant *B* be distinct from an instant *A*, then *B* will be said to be after *A*, if, and only if, it be abstractly possible for a person, at the instant *A*, to produce an effect at the instant *B*.

Robb seems here to want to side with the prevailing philosophical psychologism of the period, but, of course, the anthropocentric aspects of these glosses on *after* cannot be taken really seriously since human powers to discriminate and act are too crude to do the work required of *after*. Perhaps, also, he wished to defer to the idea that temporal relations have to be based on some other relation as a requirement of philosophical hygiene. But it is all too clear why this rather vague part of the discussion is relegated to an Introduction. Robb does not really provide a suitable basis to make plausible the claim that SR may be based on causality.

(e) Next, could we not simply include it as a postulate of SR that causality is coextensive with the light cone, since it looks highly unlikely that there are any particles such as tachyons? (This is a suggestion put to me in discussion, if I understood him, by Bas van Fraassen.) This would give us two versions of SR, a classical one in terms of which the existing tachyon literature is consistent, and a novel one which rules tachyons out simply and directly by a postulate.

That could be done, of course. But it would not hold much philosophical interest to do it. Surely what has been under close discussion in the philosophical import of SR is whether or not causality (through the Limit Principle) is the necessary foundation of the relativity of simultaneity and, more generally, of the Lorentz Invariance of the laws of physics. I have been arguing that it is not. The Limit Principle is not the basis of the Invariance Principle, which stands entirely without its support. To include the Limit Principle as a subsequent postulate may have several interesting consequences, but not the consequence that it thereby becomes necessary to the more powerful parts of the conceptual structure of SR. Its inclusion would do nothing to show us how the concepts of space and time in SR depend on the idea of causality, which it was Leibniz's programme to show.

(f) Light connections are included among causal connections in all accounts of what causality is. It is quite certain that one can define the light cone in a completely satisfactory way by means of light connection. If e_1 is in or on the light cone at e_2, then e_2 is in or on the light cone at e_1. Both relations in the disjunction are symmetrical. If e_1 is on the light cone at e_2, then the light cones at e_1 and e_2 intersect in a null geodesic, a path for a light ray connecting them. If e_1 is in the light cone of e_2, then the cone at e_1 intersects the cone at e_2 in a spherical space-like surface. For any point P on this surface there is a null geodesic linking it to e_1, and another linking it to e_2. These identify the path of a light ray linking e_1 and e_2 by an appropriate reflection at the point P on the spherical surface. It is certainly a very elementary and basic postulate or theorem of SR that *light* is not propagated outside the light cone. As just indicated, it is easy to show that every time-like connected pair of events can be connected by a reflected light ray. There can be no question at all that the relation of light connection is a conceptual and a theoretical primitive of SR. It also looks to be an ontological primitive.

Does this mean that causal theorists have overlooked a simple and eminently satisfactory version of their theory? Well, if we identify the light cone and its interior by means of light connections, it is not clear how this shows *causality* to be basic in SR. It is true that a light ray is a form of causal propagation, but SR does not seem to single out the causal aspect of light as the important feature of it. If SR entailed the Limit Principle as a basic postulate then it would be plausible to regard the fundamental role it gives to light as that of forming the limit of causality. The theory would tie light deeply into the structure of causality. But SR does not do so. It is something else specific to light which SR singles out for special notice.

It is easy to say what this feature is in the concepts of the relativity interpretation. Light is always *propagated*; it is never *at rest* in any frame, even if one admitted frames corresponding to all coordinate systems allowed by the Principle of General Covariance. That is a primary ontological fact about light. In the spacetime interpretation, it corresponds to the fact that the light cone consists in null geodesics. This requires that, in any coordinate system, both the spatial and temporal coordinate differences are non-zero for any two distinct events connected by a (non-reflected) light ray. That light divides the positive from the negative intervals of spacetime is the main theoretical feature of it in SR. It is the meet of the elsewhere with the elsewhen and this is its deep topological significance. This is the basic, structural or ontological fact about light which SR makes so prominent, and this is implied in any programme to place light as the fundamental relation of SR. The concept of light is the concept of that which is propagated. It is deeply spatiotemporal. Space and time are absorbed in spacetime by a prior absorption of each into the spacelike and the timelike (the elsewhere and the elsewhen). Light-like connections form a common boundary to these, in the sense that distinct light-like connected events are always elsewhere and elsewhen from one another. These are the deepest seated conceptual facts about the relations of the spatial and the temporal to spacetime. They place light in the conceptual position of that which travels. So the conceptual role of light in the theory cannot be divorced from these spatiotemporal ideas. It divides the time-like from the space-like. That is why light cannot be used to reduce temporal relations to causal ones.

5 Foundations for spacetime structure

The main drift of argument so far has been in this direction: the crucial task in constructing foundations for SR is to find a means of identifying the surface of the light cone and its interior. For largely philosophical reasons, causal relations have been mistakenly preferred to temporal ones as a primitive basis for this identification. The mistakes have been that SR does not postulate or even entail the Limit Principle, which is a necessary condition for causality to identify the light cone and its interior, and also that, even if the Limit Principle were true, causality turns out, on examination, not to be more primitive in SR than spatial or temporal ideas are.

I want to go on to argue, now, that, in principle, one need not base the structure of spacetime on some means of identifying the light cone and its interior. One could in principle equally well base it on some means of identifying the surface of the cone and its exterior. If one could base it, equally elegantly, on space-like rather than time-like relations then spacetime certainly would not specify causality as its *uniquely* primitive basic relation, whether or not the Limit Principle is a postulate in some formulation of SR.

There is no doubt that standard physics expositions of SR, with no philosophic axe to grind, strongly suggest some sort of priority for the time-like line in spacetime, the worldline, or at least for time-like relations and connections. How the theory suggests this and whether the suggestion springs from something fundamental in it are questions we must now pursue.

At least two features have made the idea of a worldline (that is, a material, time-like extended line in spacetime) of more interest than the counterpart idea of a material space-like extended line. Neither feature is built very deeply into SR. First, the range of familiar objects and their more fundamental spatially smaller parts readily suggests the idea of a decreasing series of spatially smaller objects with a strict point mass at its limit. This, together with the fact that familiar objects frequently endure for rather long periods, prompts the idea of an eternal point particle and of the principle of the conservation of matter. This is recognisably an approximative idea and not at all a fundamental one. Spacetime theories are much more profoundly concerned with field concepts than with proposals about the constitution of matter. The second feature is that *clocks* (of whatever kind) provide operationally for a measure of proper time, the measure of interval along timelike curves. That, too, is approximative. Real clocks may be affected by acceleration. Only ideal clocks are unaffected by the possibly intensive curvatures on a time-like curve. Finally, it is not clear that the idea of a clock is fundamental in SR.

Nothing analogous to these features can be found for space-like connections. First, space-like material lines (surfaces, volumes) lie at the limit of an increasing series of spatially larger objects and a decreasing series of temporally briefer and briefer material things. That seems to be the construction analogous to that of the worldline, at least. Nothing familiar in our experience motivates the pursuit of either series towards its limit. (The tachyon hypothesis is motivated in a quite different way, but that need not concern us now.) Secondly, though rods (ideal rods) deliver a constant rest length[3] this does not give us the direct measure of a spacetime interval. Rest length is a *spatial* measure: it tells us a distance between two parallel time-like lines (the world lines of the end points of the rod). But this spatial measure is not taken along any spacetime line or curve. The idea of a *spatial* interval (which light can travel across, for example) is best understood as tied rather to the relativity interpretation than to the spacetime interpretation. It is not an interval which connects *events*. Light can intelligibly be said to travel across *spatial* but not across space*like* intervals since these are defined as lying outside the light cone.

But these features need more careful scrutiny. We can usefully split the problem into two others. Does spacetime structure reveal a more basic or more primitive role for time or the timelike as a matter of its *formal* geometry? Does it do so *operationally*?

We can give a rather unqualified 'No' as answer to the first question. Let us look at this in the perhaps rather specialised, but also rather revealing, context of a coordinate-free metric treatment of spacetime. It will be enough to consider just the problem of defining the general spacetime interval between two point events (where the interval is not Δs but Δs^2). Standardly, the interval between any pair of spacetime points p and q is a construction using proper time and the light cone. Choose a time-like curve through p and take the points r and s in which this curve intersects the light cone through q. Let t_1 be the proper time *from p to s* and t_2 the proper time *from r to p*. (Whether the proper time is positive or negative depends on the direction in which one takes it. If time itself has a direction, this will not be arbitrary.) The spacetime interval is then, quite simply, the product of these proper times. (This assumes a $(-+++)$ signature of the metric.) It will be positive for space-like intervals, negative for time-like intervals, as in fig. 3.1.

---

3 Strictly, not even ideal rods are rigid under rotation, for example.

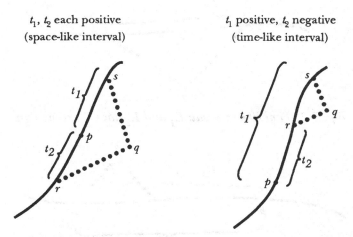

t_1, t_2 each positive
(space-like interval)

t_1 positive, t_2 negative
(time-like interval)

Fig. 3.1. Constructions like these are explained in some detail, but very simply by Geroch (1978, chapter 5, esp. pp. 80–96).

Formally, it is easy to find the dual of any such construction, using the proper length of space-like lines and the light cone. In the figures below, r and s are points in which some space-like curve through p intersects the light cone at q. Let L_1 be the proper length from p to s along the curve and L_2 the proper length from r to p. Again, the sign of the length depends on its direction. Unless space itself has a direction, this will be arbitrary. The expression for the interval is, analogously, simply the product $- (L_1 L_2)$ (for the same metric signature). Geroch also gives expressions for the spatial and temporal coordinate intervals, where the time-like line approximates the axis of a Lorentz coordinate system. The duals of these for the case where the time-like line approximates a spatial axis of a Lorentz system are given below and compared with their originals.

$$x = \frac{1}{2} (L_1 - L_2) \qquad\qquad \text{Geroch: } x = \frac{1}{2} c (t_1 + t_2)$$

$$t = \frac{1}{2} (L_1 + L_2)/c \qquad\qquad t = \frac{1}{2} (t_1 - t_2).$$

The illustration shows that space-like duals of time-like based constructions are a general formal feature of spacetime geometry. So this geometry gives no superior formal status to time-like as against spacelike connections, nor to causal relations, even if the Limit Principle is entailed by SR or a postulate of it in some axiom system for the theory.

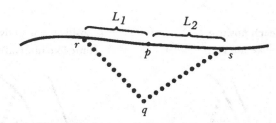

Fig. 3.2. *Time-like interval: L_1 and L_2 have the same sign.*

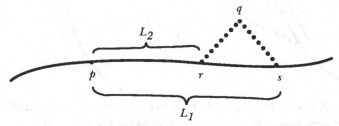

Fig. 3.3. *Space-like interval: L_1 and L_2 differ in sign.*

When we turn to reflect on operations in measurement and experiment we find what seems to be a clear preference for the time-like. We can regard an observer as *on* (approximating to) a worldline and as *sending a signal* from it to *q* and *observing the signal reflected* back from *q*. This replaces the rather abstract account of the procedure given before. The observer now takes the relevant readings of proper time directly from his clock. But no observer can be on the space-like curve through *p*, in the analogous sense. In the first of the two figures illustrating the spacelike construction, one can see how light signals can be *sent* from *q* to *r* and *s*, since the space-like line lies in the direction of the forward light cone at *q*. But if the line lies in the backward light cone at *q*, as in the second figure, it is not so clear what the procedures of signalling would be.

I doubt that we can attach much weight to any of these operational disanalogies save one. That is the lack of any space-like counterpart to a clock which would allow us to measure the interval along a space-like curve. This needs more careful pondering.

Rods deliver a constant *rest length*. From a spacetime viewpoint this is something like the constant distance between two parallel time-like lines (the worldlines of the end points of the rod). But it is not any kind of measure of a spacetime interval. Given a point on one of these parallels, infinitely many space-like lines through it and intersect the other parallel. These space-like lines differ in proper length, yet only one rest length seems to correspond to all of them. Thus the rest length seems not to give the measure of any spacetime interval. (See §4, pp. 91-4 for a fuller account.)

Nevertheless, there is (and must be) a well-defined connection between rest length and proper length, both conceptually and operationally. Take two spacetime points p and q with space-like separation. These are not the end points of a rod, of course, but they can mark events which might occur at the end points of a rod. Geometrically one takes the class of all pairs of parallel time-like lines, the first member of which contains p and the second q. Each pair represents the trajectories of the end points of a rod at rest in some Lorentz frame. Any such rod has a rest length as a constant of it. Now consider the rest lengths defined by this class of pairs of parallels. It contains a unique maximum L, together with all positive real values less than L and greater than zero. So the class defines only this maximum, L. *Now, this rest length L is also the proper length measure of the spacelike interval pq.* Every proper length may be defined by a rest length in this way.

This picture of the connection between rest length and proper length has some further interesting consequences. If we ask which pair of parallel lines establishes this necessary connection between rest length and proper length, we find the answer that it is the pair *orthogonal* to the space-like line joining p and q. That is, each parallel line is also parallel to the time-like axis of every Lorentz coordinate system for which p and q have the same time coordinate, and so they lie on the space-like axis of some such frame. In the relativity interpretation, the construction which established the necessary connection between the ideas of the proper length between two spacetime points and the rest length of rods also identifies the Lorentz frame for which two such spacetime points are simultaneous and the related rod is at rest. Thus there clearly is an essentially metrical construction which picks out a class of simultaneous events which corresponds, in a non-arbitrary way, to a given class of rest points. That is, even if we use the philosopher's idea of a frame of reference criticised for its inadequacy in §4.3, we can still associate classes of simultaneous events with it in just one non-arbitrary way; this will give us, in fact, the complete Lorentz frame of reference as a result. No assumption or connection about the speed of light is made at any point. As in Malament's undoubtedly deeper, topological construction of orthogonality, one can construct just Lorentz simultaneity or Lorentz frames and no others. It is worth pointing out that in Winnie's treatment of SR without one-way velocity assumptions (with ε treated as a free variable), expressions for both rest length and proper length contain no ε variable; that is, they are convention-free (see Winnie 1970). So this sheds further light on the claim in §3.2 that neither simultaneity nor the Invariance Principle need be conceived of as having its factual import debilitated by conventions.

Returning now to the problem of operations of measurement of spacetime, one finds that there is, after all, an operation which gives us the proper length along a space-like interval. If e_1 and e_2 occur at the ends of one's measuring rod and one sees them together (simultaneously at one place) at the mid-point of the rod, then the proper length measure of the space-like interval e_1-e_2 is the rest length measure of the rod.

It would be incautious to make a strong claim about the operational symmetry between the space-like and time-like approaches to SR on the basis of this discussion. Nor is it necessary to do so. The operations we have been scrutinising are essentially metrical ones, the question being whether the measuring of time-like trajectories is more accessible operationally than the measuring of space-like paths. But the deep and elegant studies of spacetime geometry carried out by Robb, Zeeman, Malament, Winnie and others show that a separate *metrical* development of spacetime is not necessary. What these studies make clear is that we need only procedures which will decide, for any given event, which other events are in (or on the surface of) its light cone. No metrical question need be settled. The whole metric congruence structure can be built on that very modest topological basis.

It should be easy to see, at this stage, that at least in a two-dimensional spacetime such time-like constructions as these have space-like duals. So it must also be true that for this special case we need only procedures which will decide, for any given event, which other events lie on the surface of the light cone *or outside it.* Once, again, no metrical question need arise. The whole metric congruence structure of spacetime can be built on that very modest topological basis. A construction of this kind uses the *elsewhere* relation rather than elsewhen. (Notice that, for distinct events e_1 and e_2, if e_1 is on the surface of the light cone at e_2, then e_1 is elsewhere from e_2.) So space-like relations provide a basis for spacetime structure just as satisfactory as that given by time-like ones, and so at least as satisfactory, but utterly different from, a basis which might be found in causality should the Limit Principle turn out to be a fundamental postulate of SR (despite all the evidence against this possibility). In higher-dimensional spacetimes there is always a spacelike curve from any point to any other, but not always a time-like one (a point which I owe to Adrian Heathcote). What is crucial here seems to be that time (but not space) is one-dimensional. If that does defeat the prospect of dual space-like topological constructions for general spacetimes, it is at best unclear how it rests on causality or mere time-likeness itself as a basis.

6 Conclusion

The standard view is a thesis of philosophy. It is easy enough to lose sight of this fact in the necessarily detailed reflections on SR by which someone supports or opposes the view. At present, the strongest plank in the causal theorists' platform seems to be the deep and elegant constructions by Zeeman, Malament and Winnie which show how much of the structure of SR can be built on the topology of a single relation. One needs to remember that these constructions, revealing as they are, do not themselves show more than that some one relation will do the trick. They do not show that this one relation actually is causality rather than another relation, and have, I think, no tendency whatever by themselves to settle the philosophical question as to whether it is *causality* that can or should be used to reduce temporal or spacetime relations to something else.

Another main philosophical issue on which causal theorists and others may divide is the issue of what counts as physical. I think that Grünbaum's remarks on the problem of tachyons are directed at a particular stand on this issue (1973, pp. 824–7). He speaks of 'the philosophical objections to this gambit of grafting the ontology of one theory of time order onto a physics of time order which is alien to it' (p. 824); and, later, 'Why, I ask, are the relativist time order relations still ontologically sacrosanct, if there is indeed superluminal causal connectibility among events?' (p. 827).

I take it that Grünbaum thinks that unless genidentical and causal connections are confined to the light cone then there cannot *be* anything – more explicitly, I think, there cannot be anything *physical* – to *constitute* real and therefore absolute temporal relations. But any such response depends on there being precise limits to what counts as physical and thus as the existence of physical features. It depends, I believe, on the view that time relations, not only may be, but must be *based on something else*. It depends on the view that time relations cannot themselves be physical, that spacetime itself is not an element in the ontology of physics and that features of spacetime cannot be physically basic.

This is not the place to examine the issue. (Grünbaum's writings may be seen as an extended commentary on it; my own, quite different, view is argued at length in Nerlich 1994.) But it is the same as the issue whether the Invariance Principle or only the Limit Principle can provide a properly *physical* basis for spacetime relations (quite apart from the question whether SR entails the latter). The Limit Principle sees the conical ordering of spacetime as constituted by the enclosure of matter and light within the light cone – it sees the cone as *constituted* by that limitation. That is undoubtedly physical. Indeed it is, very nearly, *material*. The Invariance Principle does something quite different. It speaks of light not as the limit of matter but as occupying a crucial structure *in spacetime*, a structure described, essentially *indirectly* by reference to a special class of coordinate systems for spacetime. Interpreted more directly, it characterises light as forming the boundary between positive and negative spacetime intervals, as the overlap of the elsewhere with the elsewhen. The philosophy which makes the Invariance Principle central to SR is a philosophy which takes spacetime and this feature of it as basically, irreducibly, physical but not (not *at all*) material. That is a main issue between distinct ontologies of the physical. Though it is correct, in some sense, to follow Minkowski in saying that, in spacetime, space and time shrink to mere shadows, it might be better to say that they are shown to be shadows of the space-like (the elsewhere) and the time-like (the elsewhen) relations. Then SR can be seen as telling us, not something about the relations of matter to light, but about the interesting union of the space-like, time-like and light-like relations in spacetime. Metaphysically, we can best understand its message, not as a materialist, but as a physicalist one about a physical particular – spacetime. This is significant not just for the metaphysics of SR but for the metaphysics of scientific theories generally. That is at least a large part of what this debate is about.

4 Simultaneity and convention in Special Relativity

1 Introduction

What physics books are apt to say about SR (Special Relativity) is not quite the same as what philosophy books are apt to say about it, as Wesley Salmon points out in his excellent *Space, Time and Motion* (1975, p. 113). He explains this difference reasonably enough, as due to disparate main interests which SR has for physicists as against philosophers. The former want to develop quickly an apparatus which allows the clear, deft portrayal of central principles and results in physical prediction and explanation. The latter prefer a more leisured approach to this goal so as to give scope for a deeper insight into the semantic-syntactic structure of SR. Most philosophy books say that the language of SR has various conventional elements in it, which means that the theory can have no very simple relation between its syntax and its semantics. In particular, the matter of the simultaneity of space-like separated events is settled conventionally, and this gives rise to a contrast in SR between sentences which form a factual core (Winnie 1970, p. 229; Salmon 1975, p. 117) and others which make up a periphery of non-factual sentences with a merely syntactic function. In what follows I ignore the problem of what other conventions might have a place in SR. I want to examine and reject just this idea that simultaneity is a convention, as this gives rise to the idea that we can contrast a core of factual sentences of SR with a periphery of merely conventional ones.

Wesley Salmon's thought on this central problem is certainly conventionalist. Besides its admirable clarity and precision, Salmon's work contains many valuable arguments and observations on the structure of SR. He has illuminatingly discussed a wide range of 'alternative methods' of synchrony (1977c). He has proposed a useful qualitative measure of conventional triviality (1969). In his book (1975) he makes explicit the very important distinction between the question whether simultaneity is a ternary empirical *relation* among two events and a frame and the question whether it is established by *convention* in a frame; he also gives an informative and provocative account there of the significance of superluminally fast particles, known as tachyons. Like many others, I have learned much from his (and other) scholarly and elegant discussions of conventionalism to which I am deeply indebted in my own attempt to state a quite different view of the status of simultaneity in SR. If it is discovered that, in learning something, I have still not learned enough, the mistakes will be all my own.

What do physics textbooks tell us about SR? The best of the more recent ones tell us that it is a theory about a four-dimensional physical manifold, spacetime, with Minkowskian metric in which we can describe a wide range of physical quantities mainly in terms of 4-vectors and tensors. In this, light propagation fills a distinguished place since electromagnetic wavefronts (in a vacuum) occupy what are called the null geodesics of spacetime. Time-like geodesics describe the trajectories of force-free particles. I assume, in common with many others, that spacetime physics, framed in concepts of the style I have just been illustrating, gives the best account of SR.

Physics books sometimes say other things, using other concepts. They tell us, for example, that whether or not two space-like separated events occur at the *same time* is a relation, not just between the two events but among the events and a *frame of reference*. They tell us that SR is special in being confined to a restricted class of frames of reference in each of which the laws of physics take the same, invariant form, and in each of which the speed of light (in a vacuum) is a universal constant *c*. It is on this latter way of picturing SR that philosophy books claim to improve. They say that, given a pair of space-like separated events and a frame of reference, it is not a relational fact but a convention whether the events are simultaneous. This means that, given both the events and the frame, we can still freely choose to say that the events are simultaneous or that they are not simultaneous. Neither choice states (or misstates) a fact and neither sentence chosen is apt for semantic appraisal nor has a truth value. We simply determine the language by a decision on the matter, the decision is conventional, and we gain no factual content for SR by making it.

I will make use of Reichenbach's ε notation. This has become familiar, but I shall briefly introduce it. Suppose there are two spatially separated points A and B in some reference frame. At t_1 on a clock at A, A sends a light signal to B which arrives there at t_2 on B's clock, is reflected back and arrives at t_3 on A's clock. Then t_2 (at B) is related to t_1 and t_3 (at A) by the formula:

$$t_2 = t_1 + \varepsilon\,(t_3 - t_1),\ 0 < \varepsilon < 1.$$

That simultaneity is conventional is equivalent to the choice of ε's being conventional.

This picture of things began with Einstein who wrote (in 1905): 'we establish *by definition* that the "time" required by light to travel from A to B equals the "time" it requires to travel from B to A', (1923, p. 40, italics in original. The translation quoted is a correction; see Grünbaum 1973, p. 344, n. 4). It is hard to overestimate the impact of this remark on philosophy in this century, set, as the words were, in the context of a highly successful, revolutionary theory of physics. Though Einstein used tools already forged by logical empiricism there is no doubt that his employment of them in this brilliant paper immeasurably enriched their apparent range and power. It is still a methodological orthodoxy in the philosophy of space and time to see many apparently real questions as calling for answer by factually arbitrary, conventional stipulations. The conviction that simultaneity questions, in particular, call for such an answer is still the central stronghold in the empire of conventionalism.

The strategy of my arguments will be clearer, I hope, if I make two comments on Einstein's predicament in 1905. Firstly, he faced a major problem in teaching us (and himself) how it is *conceivable* that simultaneity might be relative to a frame. He had to undermine what appeared to be the synthetic *a priori* truth that it is absolute. It is not surprising that he should have looked to a positivistic epistemology to solve the problem. I will argue that we can see, in hindsight, that the problem was not epistemological, and therefore that the solution was misconceived. Secondly, Einstein wrote before Minkowski discovered how to find the description of the spacetime world which lay encoded in Einstein's description in the 1905 version of SR. I shall argue that, on the one hand, a spacetime perspective enables us to see most clearly what Einstein's simultaneity problem really was, despite the fact that, on the other hand, questions of simultaneity can properly be raised and settled only in a classical ontology in which space and time, as separate entities, do *not* 'fade away into mere shadows' (Minkowski in Einstein et al. 1923, p. 75). At the same time we must provide a basis for translating, without loss of content, SR physics in its spacetime interpretation into SR physics in its space and time interpretation. A condition of adequacy which constrains the concept of a frame of reference is that *it* should provide a basis for these complete translations. Frames of reference, as conceived of by Einstein in 1905, and as generally conceived of in philosophy since, do not meet this adequacy condition, a fact which Einstein could not possibly have understood before Minkowski's great step forward to spacetime. The idea of a *convention* for simultaneity arises out of these confusions. It is to be rejected by clarifying them and cannot, I think, be shown mistaken by propounding various styles of experimental means for fixing simultaneity.

2 Coordinate systems for spacetime

I follow many others in seeing SR as primarily a theory of spacetime rather than of frames of reference for space and time. This is important since spacetime affords no natural foothold for the ideas of frame of reference, speed of signals, simultaneity of events. It is conceptually incongruous with the question conventionalism aims to settle. I shall spend a paragraph making this not unfamiliar point more explicit.

I call the language in which we speak about spacetime and spacetime objects a 4-language and the ontology of this language a 4-ontology. Now we may not say, in 4-language, that spacetime objects move in spacetime nor that they remain at rest in it. Nothing in spacetime structure entitles us to say of two time-like separated points that they are 'at' the same spatial place or 'at' different places at different times. 'Same (spatial) place' in not part of 4-language. Similarly we can neither affirm nor deny of two space-like separated points that they are 'at' the same time, or simultaneous. We cannot say of two time-slices of a spacetime object that 'it' is the same thing (at the same place again). 'It' is neither a (continuant) thing nor the same. Thus, in 4-language (and in 4-ontology, therefore) there are no (continuant) things which move or remain at rest, and nothing makes (or unmakes) the occurrence of two events be at the same or at different times. Nothing is to be said, in 4-language, about the speed of anything, not even about light's speed being (or failing to be) constant nor being the greatest speed. Light rays have a privileged status, indeed, for they lie in the cone of null geodesics which is metrically determinate at every spacetime point and distinct from every non-null geodesic of the manifold. In 4-language we can attribute invariant properties to spacetime objects; the spacetime interval between spacetime points, the shape and size of material hypervolumes in spacetime can be described invariantly, no matter by what varying coordinate differences we specify them in different systems of coordinates. Our 4-language is not a (continuant) thing language; it says nothing about a (continuant) space nor about (spatially global) times. Spacetime physics invites the use of a quite particular array of concepts to describe spacetime and its objects and claims to give an exhaustive account of these. In this array of concepts the idea of a frame of reference, in particular, has no place. I will defend this last claim in more detail in §4.3.

For describing spacetime and its physical contents it is convenient to choose a system of coordinates. This is not at all the same thing as choosing a frame of reference – not in the conventionalist's book, quite certainly, and not in mine, as will become clear. It is a truism to say that coordinates are conventional. This means at least two things: first, the quadruple of coordinate numbers ascribes no property to the point which has them in the way that a quadruple of numbers might describe an energy-momentum or electromagnetic vector, a spacetime object, at that point; second, and more importantly, we can describe the vectors, tensors, the metric of spacetime and so on without the use of coordinates, if we wish to. A coordinate system is just a global device of reference, a (dispensable) means of representation, nothing more – save for the proviso that we expect coordinate differences to reflect the smooth ordering of points along coordinate lines.

Another way to bring out the fact that coordinates are conventions of representation is to point out that we expect our spacetime physics to be formulated in a *generally covariant* way. That is, although the coordinate components of, say, a vector vary from one arbitrary system of coordinates to another we expect that the components will change according to the same transformation as allows us to go from the one set of coordinates to the other. The components are covariant with the coordinates. This effectively robs the coordinates of any significance beyond that of representation. It tells us, for example, that the privileged status of light as a null geodesic depends not at all on any system of coordinates which may be in use. (But, of course, this privileged status does not ascribe a *speed* to light, as a signal.) That SR can be expressed in a generally covariant way has been argued for in a conventionalist's paper (Giannoni 1978) and is not a matter of dispute.

I now want to make two points which seem to me very obvious ones indeed in the light of what has been said. Evidently all that a question of 'simultaneity' of two events could come to *in a coordinate system* with axes x_0, x_1, x_2, x_3 is the trivial matter whether or not the system gives them the same x_0 coordinate. Though this is indeed trivial and reflects no fact about the events and their posture in spacetime it would be quite misleading to describe it as a *convention within the system* whether they have the same coordinate or not. The system itself is a convention; we can get along without any such system. That we have chosen this system rather than another is a convention, if you like, since no fact about spacetime is reflected in choosing it. But given the system, there is no *further* matter to be settled – by convention or by anything else – as to whether the events have the same or different x_0 coordinates. That is simply a *relation* they bear to the coordinate system in use and is fixed in choosing it. Precisely parallel remarks apply to the idea of 'same place' in a coordinate system. It can come to no more than the trivial matter whether the system assigns two events the same space coordinate. But, of course, no one ever did think *that* was a convention. It was always clearly understood as relational.

Next, we need four coordinates if we are to give an adequate *coordinate* description of four-dimensional spacetime and its physical contents. That is to say, that part of the system of coordinates which assigns the same (or different) x_0 or time coordinate to space-like sep-

arated events is no less necessary for a complete coordinate description than the parts which assign the same (or different) x_1, x_2, x_3 or
space coordinates to them. Without all four coordinates for any pair
of points we cannot express, for example, the invariant spacetime
interval between the points in our coordinate language. We cannot
define which geo-desics are null by coordinate means. Each coordinate is quite as necessary as any other for the coordinate expression
of the facts which SR has to tell us about the structure of spacetime
and its contents.

Compare the coordinate geometry of the two-dimensional spatial
plane, confining ourselves to linear, but not necessarily orthogonal,
coordinates. Let us suppose that someone notes our freedom to
choose any of the lines of the plane as an *x* axis. Consequently, he says
that *x* coordinates of (and *x* coordinate differences between) points
in the plane are *relative* to a choice of *x* axis. He also notes our freedom to choose any of the lines of the plane (not parallel to the one
chosen as *x* axis) to be a *y* axis. However, he describes choice of a *y*
axis not as completing the choice of a relatum suitable for the complete coordinate description of geometrical objects in the plane, but
rather as a *convention* adopted *within* the chosen relatum of an *x* axis.
He asserts that it is a question of relational fact whether a line is or is
not parallel to the *x* axis, thus giving all the points on it the same *y*
coordinate (no matter which *y* axis we choose). Therefore, a sentence
which states this factual relationship (of parallelism) has a truth
value. However, he claims, it is something quite different, a convention, whether a line is or is not parallel to the *y* axis, thus giving all the
points on it the same *x* coordinate (whichever *x* axis we might have
chosen) so that a sentence which states *this* parallelism has no truth
value and corresponds to no state of affairs. Further, he tells us, there
are no facts about *y* coordinate cross-sections of figures generally in
the plane unless their sides are parallel to the *x* axis. Of course, there
are facts about the dimensions of these figures relative to *new x* axes,
chosen so as to parallel the sides. But relative to a given *x* axis the *y*-
cross-section geometry of figures in the plane generally is no question
of fact and can be settled only by the free conventional choice of *y*
axis. This, I suggest, would be a deeply misleading account of the
nature of coordinate geometry in the plane. It seems to me no
improvement to say that whether two points have the same *y* coordinate may be equivalently described either as a *convention within* the

relatum of the chosen *x* axis or as a relation both to the *x* axis and to the *y* axis. This appears to me to parallel Giannoni's account of the conventionalist's description of things (see Giannoni 1978, pp. 39–40).

I suggest that if anyone were to adopt a conventionalist picture of the relation of time-like to space-like axes for the coordinate geometry of spacetime it would be misleading a quite similar way. Of course, the situation is more complex in spacetime. In particular, we can define invariant intervals of spacetime without recourse to space-like axes, treating the geometry in what approaches a coordinate-free way. We can use a measure of interval along time-like lines (proper time), a 'rest length' metric which 'spaces out' parallel time-like lines (but is emphatically *not* a measure of interval along space-like coordinate lines) and the 'light ratio' which connects these metrics. This is given by what is, essentially, the familiar two-way light principle. Given two points *a* and *b* on a time-like line *A* it relates the proper time (measure of the interval) *b*–*a* to the 'rest-length' metric from *A* to any of the time-like lines parallel to it which contain just one point of the (space-like) sphere given by the intersection of the light cones through *a* and *b*. The figure (which 'doubles up' those usually drawn to represent the two way light principle) makes this graphic, I hope.

What Winnie's ε-variable treatment (Winnie 1970) shows is that provided we can measure intervals along the curves *A* we do not *also* need a measure along space-like curves (such as spatial coordinate axes for spacetime would provide). The 'rest length metric' and the light ratio suffice. I will call this a *pre-coordinate* geometry. What this means, in fig. 4.2 is that the 'rest length' tells us that the time-like

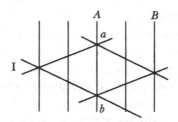

Fig. 4.1 (b–a) *measures proper time* along A *from* a *to* b. (B–A) *counts the number of unit-parallels from* B *to* A. *It is* not *a measure along any space-like curve from* A *to* B.

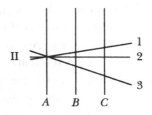

Fig. 4.2

lines *A, B, C* are, say, one rest-length metre apart without defining how this relates to the measures of the various space-like intervals between *A, B* and *C* taken along the lines 1, 2, 3.

No doubt this is deeply significant. But it does not even *appear* to deny that we *can* do full coordinate geometry if we wish. Nor does it deny that, if we do, then the selection of space-like axes closes all questions whether or not two events have the same time coordinates. Nor, again, does it seem to deny that the selection fixes a perfectly definite space-like geometry for the coordinate time-slices of 4-objects whether or not their time-like extended lines are parallel to our time-like axis, i.e., whether or not they have coordinate velocity. There are certainly factual matters about the size and shape of *all* such time-slices of 4-objects, and their inclination or lack of it to the chosen time axis is simply irrelevant. In fact, we can deduce precisely what these spatial properties are *from* Winnie's ε-treatment of SR which suffices to define the measure of *any* interval of spacetime, whether space-like or time-like. That we *need* not do coordinate geometry does not mean that when we do we find some sort of *factual slack* to be filled out by convention.

Next, it would be a mistake to suppose that we have here come across a unique geometric role for time-axes and for time-like lines

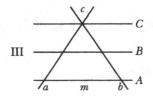

Fig. 4.3

generally. We can turn fig. 4.1 on its side (as in 4.3), make perfectly good sense of it and have it yield a route to the spacetime metric no less elegant than before. For simplicity, consider a two-dimensional spacetime. Let a and b be space-like separated points in spacetime, and let c be the point of intersection of the (upward) light cones through a and b. If we select a and b as points defining a space-like axis then a question might be said to arise whether c occurs at the mid-point (m) between a and b. Suppose someone asserts that, relative to this space-like axis it is a convention whether c occurs at the same place as m; that this can be settled only by stipulation to the effect that light from a towards b goes the same distance as it goes from b towards a in the time spaced out by the space-like parallels A, B, C between ab and c. The simplest stipulation is that c occurs at the coordinate place

$$c = a + \frac{1}{2}(b-a) \qquad \text{(compare } t_2 = t_1 + \frac{1}{2}(t_3 - t_1))$$

but so far as the physical facts are concerned, we get exactly the same physics by replacing $\frac{1}{2}$ by any other real number in the open interval $(0, 1)$. To avoid arbitrary, non-factual conventions or assumptions, we must use the expression

$$c = a + \varepsilon(b-a) \quad \text{(compare } t_2 = t_1 + \varepsilon(t_3 - t_1)).$$

Now provided we have a proper length (measure along the space-like curve A from a to b), a metric for simultaneity intervals which 'time out' the space-like parallel lines A, B, C, and the light ratio, then we can get along without a time-like axis. We can construct the geometry of space-time in a pre-coordinate way, defining the invariant spacetime intervals analogously to our (to Winnie's) procedures before. Briefly, let me show just how light-like paths define null geodesics independently of choice of ε. The space-time interval ac is given, in ε-notation thus:

$$\Delta x_\varepsilon^2 - c_\varepsilon^2 \Delta t^2 \qquad \text{(compare } \Delta x^2 - c_\varepsilon^2 \Delta t_\varepsilon^2)$$

that is,

$$(\varepsilon(b-a))^2 - \frac{\varepsilon(b-a)}{\Delta t}^2 \Delta t^2 = 0$$

$$\left(\text{compare } \Delta x^2 - \left(\frac{\Delta x^2}{\varepsilon(t_3 - t_1)}\right)^2 (\varepsilon(t_3 - t_1))^2 = 0\right).$$

I conclude that this new ε-free-variable treatment has all the advantages of Winnie's treatment and shows just as much (and as little) about what is conventional and what relational in SR.

Admittedly there is no simple means of measuring the interval along a space-like track analogous to the way a clock measures a time-like one. But this is beside any point about which sentences are factual and which conventional. First, we can measure the interval operationally by taking the (unique) maximum of the rest length measures between all pairs of parallel time-like lines which include points *a* and *b* respectively. This assumes nothing about simultaneity nor does it need the passing of signals. There is no way to reject the identity between this maximum of the rest length metrics and the proper length along the space-like interval while keeping intact the links which bind the metric of spacetime to natural measures of space and time. Again, Winnie's ε-variable treatment allows us to deduce the proper length (and, indeed, had better do so on pain of inadequacy) since it is a spacetime invariant. It is worth noticing that the pair of parallels which gives us this maximum is, in fact, orthogonal to the space-like line *a*, *b*, thus parallel to the time-like axis of any Lorentz coordinate system which takes the line as its space-like axis. This appears already to provide us with a signal- and transport-free criterion for preferring $\varepsilon = \frac{1}{2}$.

There may indeed be much interest in pursuing SR without assumptions as to the distance-in-one-direction-per-unit-time with respect to space-like frames of reference. But it seems to me quite gratuitous, indeed actively confusing, to regard rest or motion not as relative but as conventional for this reason. All the expressions, in my attempt to paraphrase Winnie's treatment, which concern *same place, rest, motion*, and so on are quite incongruous with what is under discussion and cloud a clear insight into what it is about. Questions of rest and motion, like those of simultaneity, have no place in spacetime geometry.

Despite the much greater complexity of spacetime geometry then, it does seem that a conventionalist style of treatment of coordinate geometry of spacetime is no less misleading than the conventionalist account of the coordinate geometry of the Euclidean plane, which I parodied before. It is true that choosing a time-axis leaves us perfectly free to choose spatial axes from among any of the space-ike lines of spacetime. But it is equally true that choosing a set of three spatial axes leaves us quite free in the choice of time-axis. General covariance guarantees that. There is no priority among axes here despite the fact that we may not rotate any time-like axis into a space-like axis (though we may rotate any space-like axis into any other). From the standpoint of spacetime physics one finds no reason at all for singling out x_0 or time-axes for special status. They appear to have no sort of superior philosophical interest, nor do spatial axes create a special foothold for the use of conventions within coordinate systems. Difference choices of axes in no way disturb the privileged status of the light cone as composed of null geodesics, nor do they affect in the slightest the invariant intervals of spacetime itself. In fact, from this viewpoint it seems impossible to discover anything which conventionalist theories of simultaneity shed the least light upon.

3 Frames of reference

However, these are merely manoeuvres preliminary to fixing on the area where I take conventionalism to be properly debatable. Conventionalists argue not about coordinate systems but about frames of reference. My next objective is to make clear what conventionalists have meant frames of reference to be and that much of the difficulty about simultaneity has been created by this idea's being inadequate and unfruitful as conventionalists apply it. It is clear enough that some epistemological advantages make philosophers' frames of reference seem important, but I will argue that the less said about them (and expressed by means of them) the better. Unless, of course, we give up talking about the sorts of frames of reference philosophers usually deal with, and follow the usage of physicists by adopting a fundamentally different concept of frame of reference.

A frame of reference is something which picks out of spacetime a space and a time. This is rough and a bit misleading, as we will see, but it does suggest at once why a frame of reference is not just a coordinate system. A frame of reference is insensitive to many criteria which distinguish coordinate systems, such as the difference between polar and Cartesian coordinates for space, a rotation of spatial axes and a translation of the origin in space or in spacetime. There is a 1–1 correspondence between frames of reference and *classes* (uncount-

3 Frames of reference   103

ably large classes) of coordinate systems. Members of one of these classes are alike in respects more deeply significant than those which distinguish one member from another. What is the significance of this? It is the importance of a point of transition from one array of concepts for dealing with physics to another array; it is the significance of a bridge towards an alternative language and ontology.

A frame of reference only *corresponds* to a class of coordinate systems. For example each continuous set of spacetime points all with the same x_0 (time) coordinate in some system of coordinates picks out a space-like hypersurface in spacetime. But none of these hypersurfaces is a space since the space-like hypersurface $x_0 = t$ is *not identical* with the hypersurface $x_0 = t_1$, and no point in the one hypersurface is the *same place* as any point in the other. Unlike any coordinate system, a frame of reference enables us to speak of space, of time, of continuant things with a three-dimensional shape and size which endure in time, of rest (the *same* place at *different* times), of motion, of speeds and velocities, and, lastly, of simultaneity (the *same* time at *different* places). These, like the term 'frame of reference' occur not in 4-language but in another language, with other conceptual structures and with another ontology. It is a space and time, or a 3+1-language (and ontology). The idea of a frame of reference is much older and more familiar than that of a spacetime coordinate system, deriving from classical, post-Newtonian mechanics. It gives point to a number of questions which have only trivial counterparts in the ontology of spacetime. To be specific, it gives them an ontic or structural point. In 3+1-language we can pointfully ask whether the speed of light is the same in all directions, what the (spatial) shape and size of an object is, whether it is moving or not, which way and how fast. Indeed, the syntactic structure of the language *requires* that we ask these things.

It is unclear how seriously we can or should take the metaphysical view that the 3+1-physics is no trivial alternative ontology for SR. But unless it is a significant alternative then questions of simultaneity, the speed of light and the rest are not significant either. That the ontologies differ only trivially would be no ground for recommending conventionalism but rather for seeing matters in the light of the observations on spacetime coordinate systems made in §4.2. These make no sort of sense of 'conventions of simultaneity'. Unless 3+1-physics provides a significantly different point for questions which have no point in 4-physics then conventionalism has nothing to offer but a distinctly confused account of spacetime. I incline rather strongly toward the view that the ontic shift is significant (though *my* inclination is not really relevant to the arguments of this paper). The *motivation* for a 3+1-ontology is clear enough. It provides the concepts under which SR is epistemically accessible to us. However, the conventionalist doctrine is about the factual content of SR as contrasted with its linguistic content. It claims that SR has a factual core and a merely linguistic periphery. (See, e.g., Salmon 1975, p. 117. §§1.7 and 2.5 of this book expand this core–periphery picture of conventionalism and criticise it in connection with affine and metric properties.)

To make clear how a 3+1-ontology might give importance to questions which have no corresponding point in 4-ontology I will look at an example which I think begs no question against ε-conventionalism. Consider the problem of the relativity of motion. In classical physics we cannot take as at rest, frames accelerated or rotating relative to inertial frames since this creates, quite artificially, inertial, centrifugal or Coriolis forces of which we can give no proper account. We would be obliged to posit *causes*, yet there are none. However, consider generally covariant formulations of SR. I mean by this that we deal with flat spacetime but place no restrictions on coordinate systems that they be linear. In some usages this formulation is regarded as the beginning of General Relativity, whereas in others this is seen as beginning only with the transition to spacetimes not necessarily flat. Then we can formulate our 4-physics with respect to arbitrary curvilinear coordinate systems. If we now treat frames of reference as trivial equivalents of these coordinate systems, then we get in some sense, a physics in which the previously unacceptable accelerated or rotating frames are recognised as proper. Of course, we get the artificial inertial, centrifugal and Coriolis forces exactly as we do in classical physics. There is no better explanation available for them than there was before, yet it seems no less pertinent to ask for their causes and sources provided that we take the 3+1-ontology seriously. Now, however, we simply wave these objections away as trivial since we refuse to recognise a significant difference between coordinate systems and frames. In effect, *we reject Newton's ontology* as a serious alternative to the new 4-ontology but continue to pay lip service to classical language in the insignificant forms of translation just described.

If the idea of a frame of reference is to lend some significance to the debate of whether simultaneity is a convention as against a relational fact, then a frame of reference ought to be a bridge between the ontology and ideology (array of concepts) congruous with space-time and the ontology and ideology congruous with a (continuant) space which endures in time. At least two criteria for the adequacy of candidates for frames of reference seem to be indicated by this:

(*a*) A frame of reference maximises the class of well-formed sentences in 3+1-language which can be semantically appraised (state facts) in terms of structures having well-defined counterparts in 4-ontology (and vice-versa).

(*b*) There should be no *facts* stateable in 3+1-language which call for (causal) explanations which the resulting 3+1-physics cannot provide.

These criteria are loosely stated but it would be laborious to make them more precise and they will serve my purpose. They beg no question against conventionalism as far as I can see. What are candidate correspondences between reference frames and classes of coordinate systems?

Let us define an *L-frame of reference* (*or linear frame*) as a class of linear spacetime coordinate systems equivalent save for translations of origin in spacetime and for rotation of any x_i coordinate about any x_j ($i, j \neq 0$). In the correspondence, an *L*-frame obliterates much that serves to distinguish coordinate systems (position of origin, orientation of spatial axes one to another) but it is sensitive to which partitioning into a space and a time is brought about by a given coordinate system. It disallows any tilting of spatial coordinates (spacelike coordinate hyperplanes) about the temporal and vice versa. The class of coordinate systems gives us a family of parallel time-like lines, each having constant x_1, x_2, x_3 coordinates in any member of the equivalence class, and each having a definite proper-time metric along it. The class also gives us a definite family of space-like hypersurfaces, each point in any hypersurface having constant x_0 coordinates in any coordinate system in the equivalence class and each having a definite (Euclidean) metric. Thus, for each point event in spacetime an *L*-frame provides a corresponding (spatial) place of the event at other times and a corresponding time of the event at other places.

The Lorentz frames of the physicists are one, but only one, kind of linear frame. In the present section it is not my purpose to consider whether Lorentz frames form a preferred class of linear frames (which is, more or less, the question whether $\frac{1}{2}$ is a preferred value for ε). That is the topic of §4.5. My present aim is to contrast L-frames with the (philosophers') T-frames so as to consider the question of which matters in SR are factual and which conventional (which is, more or less, the question what is convention and what fact when we assign ε *any* value we like in the open interval $(0, 1)$). I shall argue that to say there are conventions at issue is to misdescribe the way in which 3+1-sentences correspond to 4-sentences.

If our main requirement on a reference frame is that it should bridge the two ontologies so that we can find the same physics in both, the L-frames go a long way toward meeting it. Sentences in the language of the 4-ontology correspond quite unambiguously to sentences in the other language, and vice versa. Each L-frame has a continuant space and we can speak of the same place at different times so that description of the relative inclination of timelike lines in spacetime corresponds in a definite way to a description of rest or motion of particles. The space of any frame has a well-defined Euclidean geometry, and each continuant object of the 3+1-ontology has a definite shape, size, relativistic mass and so on whenever the counterpart object in the 4-ontology has well-defined counterpart properties. Further, if we look into any of the (infinitely many) coordinate systems which are members of the frame we do not thereby discover new properties fixed by these systems which we would be obliged to recognise as physically significant in the 3+1-ontology.

If we see frames of reference in this ontological light, then they are not conventional in the same sense as coordinate systems are, despite the close connection between coordinate systems and frames. We can do 4-ontology physics without coordinate systems but we cannot do 3+1-physics without frames of reference, since statements about the facts of rest and motion cannot be made at all except *in relation* to some frame or other. There is no 'frame-free' 3+1-physics, however free we may find ourselves to choose among frames. Further, we would certainly expect an adequate bridge between the 4 and 3+1-languages to provide for an adequate semantic appraisal of all the properties which the syntactic structure of 3+1-language provides for things. For example, there are going to be well-formed sentences attributing a definite metric shape and size to every object. If, under any proposed correspondence between the languages, the well-

formed sentences cannot be semantically appraised then, on the face
of it, that is a reproach to the correspondence rather than an indication that we are to eke out the correspondence by arbitrary assign-
ment of 'truth' values to certain of these sentences. Certainly if
another form of correspondence *does* fully provide for the semantic
appraisal of the sentences, that surely means, not that *conventions* are
appropriate in the proposed correspondence, but rather that it is an
inappropriate form of correspondence.

Let us focus, now, on the main problem we are pursuing, as it
applies to *L*-frames. *L*-frames define *same place.* Do they define *same
time* or simultaneity relative to an *L*-frame? Quite obviously, they do.
Given an *L*-frame, there is no room whatever for a convention to settle whether events e_1 and e_2 are simultaneous: there are no syntactically different, semantically equivalent 'redescriptions', no alternative
choices of e. Simultaneity is relative to a frame, however, there being
no absolute space or time common to *all* such frames. That simultaneity is well defined for a linear frame is the secret of this object's
success in bridging the two ontologies.

Plainly, this is not the idea of a frame of reference which we find in
philosophy books.

What, then, is the conventionalist's concept of frame? It is the idea
of a *T*-frame (or timelike frame). A *T*-frame corresponds to a class of
linear space-time coordinate systems equivalent save for translation of
origin in spacetime and for tilting of any x_i coordinate axis about any
x_j axis ($i = 1, 2, 3; j = 0, 1, 2, 3$). *T*-frames, then, correspond to classes
of linear coordinate systems equivalent just in that their x_1, x_2, x_3 =
constant timelike trajectories yield the same family of parallel time-

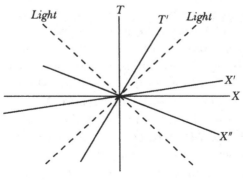

Fig. 4.4.

like lines. A *T*-frame does not define a partitioning of spacetime into space-like hypersurfaces. Each *T*-frame contains *all* the families of space-like hypersurfaces into which we might partition spacetime in the sense that it is the set of all affine coordinate systems with the same time-axis, no matter how the space-like axes of these various systems may be disposed. In fig. 4.4, each of the coordinate systems *Tx*, *Tx'*, *Tx''*, *T'x*, *T'x'*, *T'x''* is a member of a different *L*-frame from the others. But *Tx*, *Tx'*, *Tx''* are all members of the same *T*-frame which is defined just by *T*, whereas *T'x*, *T'x'*, *T'x''* are all members of the different *T*-frame, defined by *T'*.

More illuminatingly, perhaps, a *T*-frame is best understood not as the 3+1-counterpart of a class of coordinate systems for spacetime but rather as the counterpart of what I called a pre-coordinate system (in §4.2). As we saw, this treatment of spacetime is not really a coordinate geometry description, though certainly a no less satisfactory one. The question whether *T*-frames provide a suitable basis for a correspondence between 3+1- and 4-languages need not be at all the same as the question whether pre-coordinate procedures for describing spacetime geometry are suitable. We already saw that there cannot be a frame-free treatment of 3+1-physics.

It is reasonably clear that those who have debated the issue of conventionalism in simultaneity have intended *T*-frames to be frames of reference. After all, *L*-frames obviously do not permit a convention, whereas it might seem that *T*-frames do. What do we find in the literature? Frequently the phrase 'frame of reference' is used without any gloss at all. Einstein spoke, in 1905 of 'a "stationary" coordinate system' (1923, p. 38) and he is followed in this by Reichenbach, who also speaks of the relativity of *motion of bodies* (1958, section 34, esp. p. 217). This suggests that a body or set of rest bodies would constitute a frame, which would be equivalent to specifying a *T*-frame. Of course one finds many discussions of the distinction between inertial and non-inertial frames of reference, but these focus on questions of dynamics and very seldom venture anything which clarifies explicitly which *spacetime* objects are under discussion. All of this permits the *T*-frame interpretation, though one could wish for something more specific. Janis (1969, p. 74) defines a frame of reference as a family of time-like curves with spatial coordinates attached and a metric along these curves fixed by proper time intervals. That seems quite clearly to be a pre-coordinate system and plainly corresponds to a *T*-frame.

So far as I can discover, only van Fraassen uses the phrase to refer to a coordinate system for spacetime. This puts him in the position of claiming that a frame of reference is *constituted* by conventions (1969, p. 68). He is not quite so explicit as this might suggest but, on my understanding, is firmly committed to the claim attributed to him here. It allows him to argue, in effect, that simultaneity is among the conventions which constitute a frame but it does not allow him to say (nor does he say) that it is a convention within a frame. I think that this difference between van Fraassen and others is not of great significance, in the last resort. Giannoni (1978, pp. 22–5) clearly equates frames of reference with *T*-frames. (Incidentally, Giannoni, like many others, speaks of 'velocities for light . . . along the *x* axis' (p. 20) without reminding the reader that this is not at all the same thing as the *x* axis of any *spacetime* coordinate system. Perhaps this is obvious, but I, at least, would feel that I understood conventionalism more clearly if the distinction between the rest points of the *frame* and the *x* axis of a *coordinate system* were always sharply drawn. It would be strictly nonsensical to speak (*a*) of light's *moving* (*b*) as along the *space-like* line of an *x* coordinate axis of spacetime.) I take Winnie's ε *variable* (1970) as ranging over all the *L*-frames in a *T*-frame. This crucial observation has already been made (not in quite these terms) in Friedman (1977), p. 419.

It is quite clear that we cannot say, of any pair of space-like separated events, either that they are or that they are not simultaneous relative to a *T*-frame. However, it would be a confusion to think, therefore, that simultaneity needs to be fixed *by convention* in a *T*-frame. Simultaneity sentences are conceptually incongruous with *T*-frames as relata. More clearly and explicitly, *T*-frames fail to provide a basis by means of which well-formed sentences necessarily included in the syntactic resources of 3+1-language can be determinately matched with sentences in 4-language. Sentences about simultaneity are only one of the kinds of sentences which *T*-frames do not allow us to map. Therefore, *T*-frames fail a main requirement placed by SR on the idea of a frame of reference. This may not be obvious. After all, we can speak of the same place at different times, that is, of rest and motion in a *T*-frame. Further, each *T*-frame appears to identify a continuant space which has a definite (Euclidean) geometry. More, objects at rest have well-defined physical properties (shape, size, temperature) each of which corresponds to a physical property of its 4-dimensional counterpart. But objects in motion relative to a *T*-frame cannot be said to have any of these latter properties. A *T*-frame provides a determinate mapping for (sentences expressing) these properties only up to affine versions of them, so that we cannot even specify their size relation to one another or to rest objects. Salmon (1975, pp. 87–9) gives a striking example of the kind of indefiniteness I am speaking of. Whether the moving train is wholly inside the tunnel or overlaps it on either side is indeterminate in a *T*-frame. Yet the corresponding (coordinate) properties of the 4-ontology counterparts of these objects are no less well defined than they are for the 3+1-objects at rest relative to the *T*-frame. Though certainly well defined, they are not defined in quite the same way. The length of a rest object is defined in spacetime by the 'rest length' spacing out of time-like parallels, whereas the length of a moving object can only be defined as the length of a *space-like interval* of spacetime. But this difference is not a difference in any *determinacy* of factual structure. Both lengths can be expressed in the full coordinate geometry of spacetime and the pre-coordinate approach discussed in §4.2. In fact, a whole range of Lorentz-variant properties of 3+1-objects are ill-defined in a *T*-frame for any object in motion relative to the frame. However there surely must *be* definite properties of these moving 3+1-entities. If not, the counterpart 4-entities would be indefinite, too. *T*-frames in short do not allow us to construct the correspondence need to provide a semantic role for an important class of 3+1-sentences, though a semantic role *can* be provided by using *L*-frames as a bridge. So, using *T*-frames as a bridge, the 3+1-theory says less than it might, and it does not map the 4-language counterpart facts fully. It is perfectly clear what to conclude. *Choice of a T-frame does not allow us to state the full physical content of spacetime physics in the language of a 3+1-ontology.*

There appears to be something like agreement among conventionalists that to choose a *T*-frame leaves a need of some sort for a specification of ε. But is this a convention? First, choice of ε adds to the *class of sentences in* 3+1-*language which can be appraised semantically.* That is, it increases the number of well-formed sentences in 3+1-language which correspond to facts which have well-defined counterparts in structures of 4-ontology. I do not see how this can fail to mean that it adds to the factual content of the corresponding 3+1-language. Choice of ε does not *result* in there being well-formed 3+1-sentences which lack truth, state no fact, have no semantic role in SR.

Exactly the opposite is true. Well-formed sentences which are *required* (by syntactic structure) to state facts if the 3+1-language is to correspond everywhere to 4-language are not yet provided with a fact to state, with a basis for semantic appraisal, so long as ε is unspecified. In particular, facts about space-like coordinate cross-sections of objects whose time-like lines are not parallel to the time-axis are just as real and objective, just as determinate, as facts about the cross-sections of objects with no coordinate velocity. To put the point a little differently, if we discriminate more finely among the coordinate systems of those classes each of which corresponds to a *T*-frame we do thereby discover new structures defined by these systems which provide facts for 3+1-sentence to correspond to. Hence the virtue of *L*-frames.

I am arguing then that the *full coordinate geometry* of spacetime forms the most satisfactory basis on the side of 4-ontology for an SR correspondence with 3+1-ontology and that, therefore, an *L*-frame forms the most satisfactory basis on the 3+1-side. In that case, simultaneity, like rest, is simply relative to a frame of reference of the kind appropriate to 3+1-physics. The correspondence between an *L*-frame and a pre-coordinate description of spacetime is not satisfactory. This is reflected in the emergence of the ε-problem and the uncertain doctrine of conventionalism which follows it. The need for *L*-frames rather than *T*-frames is not well reflected in the view that ε ought to be specified as a convention though it does not matter how. Nor do I think it illuminates the matter to regard the conventionalist theory of ε-choice as equivalent to providing a freely chosen *further term* for a relational fact. That is, to equate for example the claim *that 'e$_1$ is simultaneous with e$_2$ relative to F' is not true but conventional* with the claim *that 'e$_1$ is simultaneous with e$_2$ relative to F and to ε = $\frac{1}{3}$' is factually true*. This is quite explicitly Giannoni's view. See Giannoni (1978), pp. 39-40. First, as was argued in §4.2, we can use pre-coordinate methods for space-time just as effectively through *space-like* lines, planes and hyperplanes as through time-like lines. It would be no less an error here than in spacetime to see the structure of SR conferring superior ontic status on time-like as against space-like elements. Second, it seems superfluous to see the need for proper semantic appraisal of 3+1-sentences as calling for *two* relata, as *T*-frame and ε, when the concept of an *L*-frame as a *single* relatum is so natural and so readily defined.

One might look to find an asymmetry which favours *T*-frames over

L-frames in the epistemology of SR which, after all, is what motivates serious attention to 3+1-ontology. There can be no doubt that our experience makes the distinction of rest and uniform motion in a frame much more accessible to us than is the shape of a rapidly moving body. This is a large problem on which I shall make only three brief comments. First, I take ε-*conventionalism* not to be an epistemologically based theory but to be concerned with the structure of the real world as SR describes it. Secondly, though the facts of rest and motion are more directly accessible epistemologically than facts about the space-like cross-sections of bodies inclined (in spacetime) towards our *x* axis, this by no means implies that the latter are inaccessible. They are in fact deducible quite unambiguously by the pre-coordinate methods discussed in §4.2. Lastly, SR neither states nor presupposes that the world fails to provide us with directly accessible, naturally occurring spacelike hypersurfaces. In fact, the world does not provide them. But suppose that a continuous three-dimensional subset of spacetime points were all to show a particular feature (radiation of a certain frequency is emitted from each, say). If this subset of points constitutes a space-like hypersurface it would certainly be quite directly observable that it did. There might be regularly 'spaced out' series of these, each member parallel to the others (analogous to parallel time-like lines), members of any one series being inclined in spacetime to members of others. Then the world would present us with choices between simultaneity classes just as accessible observationally, just as intelligible and just as purely relative as it now presents in respect of classes of rest points. So far as I understand what might be involved in this (and this proviso might be important) I do not believe that SR tells us *anything at all* about why the world is not so structured. The physics of SR is at least as much a physics of fields as it is of particles – in fact there are no strict particles in reality – and nothing about its structure, expressed either in 4-language or 3+1-language, provides a necessarily superior ontic or structural status to time-like lines as against space-like ones.

4 Historical remarks on the concept of a frame of reference

A very brief historical sketch clarifies, at least to some extent, why *T*-frames have assumed an importance which they really do not have. It also helps to make it intelligible why a disproportionate emphasis has been placed on the claim that the speed of light is the *fastest* signal and I hope to use this material to show how the conventionalist picture of SR leads to misunderstanding of the theory.

Newton understood motion as the relation of a body to an absolute (unique) continuant space. It was motion *through* space. He conceded that there might be no body at all in a state of rest, i.e., remaining in the same place in this unique space, and he also points out (roughly following Galileo) that a state of rest was not distinguishable by experiment from uniform linear motion through space. This led later physicists to regard it as pointless to attempt the discovery of Newton's unique space and to construct frameworks suitable for the conduct of physics by letting the laws themselves delete 'rotating' or 'accelerating' frames in favour of inertial ones. Had it been true that choice of a framework could not influence which classes of events are simultaneous it would have been true that choice of a suitable framework was equivalent to choosing some body or bodies as at rest. This presupposes that there are likely to be bodies which can be taken as at rest in inertial frames. However, it is not likely at all, and the presupposition has made the relativity of motion look much more like a principle which a positivist could embrace with a clear epistemological conscience than it has any right to look. From the point of view of a Newtonian spacetime, choosing rest points is equivalent to choosing a family of parallel time-like lines, but since the stacking of spatial hypersurfaces is unique up to sliding and shearing transformations (affine transformations) such a framework does provide a basis for translating 4-sentences of classical physics into 3+1-sentences without loss of content, without rendering well-formed sentences impossible to appraise semantically.

In the light of this, it seems correct to say that, in 1905, Einstein's real discovery was that the laws of physics take the same form under Lorentz transformations from any one *suitable basis* for physical description to any other *suitable basis*. This statement anticipates (and depends on) later arguments (§4.5) that, among linear frames of reference, only Lorentz frames are suitable. At present, I have argued only that *T*-frames are not suitable. Part of this discovery is that not all speeds vary under such transformations but that a certain finite speed (namely the speed of electromagnetic propagation in a vacuum) is an invariant of the transformations. This is how Minkowski described the discovery in 1908 when he gave it a spacetime interpretation. (Einstein et al. 1923, pp. 78–9). It asks us to recognise that the classical means of identifying a frame of reference, by choosing a family of rest points, no longer provides a suitable basis for 3+1-physical description. I am arguing that Einstein misunderstood this aspect of his discovery and that later conventionalists have persisted in the misunderstanding.

The account just given is, I argue, more accurate than the account which tells us that his main discovery was that light is the first or fastest signal. The latter account suggests both that Einstein's epistemological style of treatment was proper, that many 3+1-sentences of SR physics have no truth value, and that admitting 'particles' moving relative to some proper frame of reference at 'speeds' greater than light would undermine SR. Had this latter account been correct, then Einstein's discovery would have been of even more fundamental importance than it was, yet very difficult indeed to assimilate into physics. It would have been the discovery that objects in motion are *indefinite*, that there is no matter of fact about what shape or size they have and that though we attribute shapes, sizes, and much else to them in our 3+1-physics, such sentences merely are conventional, have no truth value, state no fact, are not suitable objects of semantic appraisal. However, this lack of definition would, so to speak, be movable from one set of objects to others as we change our frame of reference, in a way which seems distinctly bizarre. If this were the structure of the world it is not clear either why we should pretend (by adopting an arbitrary convention) that moving objects are definite in ways in which they are actually indefinite, why we should want to reify, on this non-factual basis, the whole new ontology of spacetime, or alternatively, how we could understand a physics in which the content of perfectly factual sentences in the language of 4-ontology becomes fugitive, conventional and indeterminate in the allegedly equivalent language of 3+1-ontology. In the event of this last outcome we would surely be obliged to conclude that 4-ontology provides, not just a more perspicuous, but the only possible ontology for SR. Surely it is clear that once we take seriously a 3+1-ontology, then there *must*, in fact, be definite shapes, sizes and so on which moving bodies have just as much as rest ones do and that this demands a simultaneity relation which is itself a *factual relation* on which all these other factual relations are dependent.

This completes the longest part of my argument. Let me state, again, what I think I have established. The major error in conventionalism has been to attempt to give an account of SR based on an inadequate 3+1-concept of a frame of reference. Unhappily most opponents of conventionalism have paid no closer attention to frames than the conventionalists themselves so that the status of those sentences left, so to speak, with no means of semantic support has been obscured. I hope that my view of their status is now quite clear. Unless every last one of them states a fact, then the content of 3+1-theory is not equivalent to that of 4-theory. This certainly does not point to the need for a convention which retains these sentences without giving them a semantic role. On the contrary, it points to the inadequacy of philosophers' frames of reference as a basis for complete description of the facts which SR tells us make up the world. The first lesson is: drop *T*-frames and use *L*-frames instead. The second lesson is: relative to *L*-frames, sentences about simultaneity, the speed of light and the rest are factual, informative about physical reality, true or false. That is the pivot round which the rest of this paper turns, if it turns at all.

5 The privileged status of Lorentz frames

A crucial problem still lies before us. Lorentz frames are a proper subset of *L*-frames. Are they a privileged subset or can we equally well choose any linear frame for our 3+1-physics? From a more physical point of view this might look like the very same question as the one we raised at first: can we choose any value for ε in the open interval $(0, 1)$? However, on a hopeful view of the arguments of §§4.2–3, it has now been stripped of its conventionalist associations and is seen as a question about the *relativity* of simultaneity among *L*-frames just as surely as questions about rest and motion are not conventional, but relational. We need only bear in mind that simultaneity, the shape and the size of moving objects are relational, factual questions to see that the facts which lead us to prefer Lorentz frames are quite obvious and simple ones.

First, there are the facts about the slow transport of clocks, already sufficiently discussed in Ellis and Bowman (1967), Ellis (1971) and Fried-man (1977).

Second, as Malament has shown (1977b), the class of Lorentz

frames is uniquely definable in spacetime in terms of the concepts of time-like, light-like and space-like connectibility. Malament says (p. 293) that this definition is provided in terms of causality, but I suggest that the description given here is more satisfactory in view of my claim that time-like lines have no privileged status in SR. This is a result of very considerable importance, I believe. Nevertheless one might appeal against it on behalf of *L*-frames by pointing out that, if spacetime really is an entity in its own right, then there can be no preferred *coordinate system* (or set of systems) which is to be used in describing physical events in that manifold. This simply enunciates a principle of general covariance for real physical entities in a real spacetime. However, as was argued earlier (pp. 103–5), questions which have no point in 4-language may well, necessarily, have point in 3+1-language. Our second adequacy condition for a frame of reference, stated there, was as follows: 'There should be no facts statable in 3+1-language which call for (causal) explanations which the resulting 3+1-physics cannot provide.' We can exploit this to provide other grounds for viewing Lorentz frames as privileged.

There is a quite clear and simple reason for preferring Lorentz frames, since only these provide us with an isotropic space for the frame. Clearly, space will not be isotropic for the propagation of light in any frame where $\varepsilon \neq \frac{1}{2}$. But not only this is the case. The length of a moving object, the rate of a moving clock, relativist mass, and so on will all be functions, not just of speed, but also of direction of motion in any *L*-frame when $\varepsilon \neq \frac{1}{2}$. In fact, once again, for a wide range of physical properties of 3+1-objects which vary under Lorentz transformation, there will be dependence not just on speed, but on direction of motion. Consequently there will be facts statable in 3+1-language which call loudly for causal explanation, *once they are recognised as* (relational) *facts*. They would require causal explanation for reasons quite similar to those which lead us to require explanation of Coriolis forces etc. and to reject certain (rotating) frames of reference which allow for no such explanations. The resulting 3+1-physics certainly cannot explain them.

Just such a criterion for preferring Lorentz frames was discussed by Grünbaum (1973, p. 355). I have not found any later discussion of it. His dismissal of the criterion relies heavily on the conventionalist philosophy which the arguments of §§4.2–3 were directed against.

The contention that either the isotropy of space or Occam's 'razor' is relevant here is profoundly in error, and its advocacy arises from a failure of understand the import of Einstein's statement that 'we establish by *definition* that the "time" required by light to travel from *A* to *B* equals the "time" it requires to travel from *B* to *A*.' For, in the first place, since no statement concerning a one-way transit time or one-way velocity derives its meaning from mere facts but also requires a prior *stipulation* of the criterion of clock synchronization, a choice of $\varepsilon \neq \frac{1}{2}$, which renders the transit times (velocities) of light in opposite directions *un*equal, cannot possibly conflict with such physical isotropies and symmetries as prevail *independently* of our descriptive conventions. (1973, p. 355)

Grünbaum is surely correct in arguing that *if* one accepts that simultaneity is not a *relation* to a frame of reference but a *convention* within it, then one must concede that further apparent facts which would determine simultaneity must be conventional, not relational, too. In that case, the isotropy of space would be a convention, not a relational fact. But, I have tried to point out, simultaneity *is* relational, not conventional, so the anisotropy of the space in the frame is real, not merely embodied in a sentence which lacks truth value, states no fact, cannot be appraised semantically. The criterion is formulated for the 3+1-ontology, not for spacetime; it yields a non-arbitrary relational simultaneity but not an absolute one. Since the criterion meets the conditions of adequacy argued for earlier, I conclude that Lorentz frames are to be preferred in SR, not just on grounds of simplicity, but because they alone achieve a partitioning of spacetime into times and isotropic relational spaces.[1]

Once we allow the simultaneity of spacelike separated events to be a matter of fact relative to a suitable frame (i.e., a relational fact), the physical *evidence* to support standard signal synchrony as physically

---

1 As I understand Janis (1983), he argues from a generally covariant expression of SR to the acceptability of skew frames of reference, thence to the conventionality of the choice among definitions of simultaneity. But this fails to make the pertinent philosophical distinctions developed here.

correct is truly massive. It, uniquely, makes space isotropic. It makes the lengths of rods, the rates of clocks, relative mass and so on, functions of a motion's speed but not direction. It gives light the same speed in all directions. It is verified by the slow transport of clocks. It is not easy to think of other matters of relational fact for which one can assemble so impressive a body of confirmation.

This completes my arguments for the conclusion that Lorentz frames give, uniquely, a factually satisfactory picture of SR in 3+1-ontology.[2]

2 I am indebted to Chris Mortensen for his careful discussions with me about this paper.

5 Motion and change of distance

1 Introduction

Sometimes things move; sometimes they stay in the same place. These are old, simple and primitive thoughts. Even so, they pervade the most subtle and sophisticated modern physical theories. I will discuss the impact of relativity theory on these old simple ideas. I hope to highlight some simple, basic presuppositions in classical thinking about rest and motion that have turned out false and which are still widely thought to be obvious. I have to say a little about coordinate systems for spacetimes or frames of reference corresponding to them. But I look at them only to draw some lessons that might be applied to the older, simpler ideas. I try to give a rational reconstruction of our simplest ideas of rest and motion; so I offer speculation rather than proof. Though I am guessing, I hope the guess is educated enough to shed some light and hold some interest.

I shall begin with John Locke, an author whose thought has much influenced C. B. Martin. Martin taught me to admire Locke and his virtues of realism, or rather, perhaps, of ontic seriousness. If this paper succeeds it will have something of those virtues and reflect Charlie Martin's benign influence, as so much of the writing of his students does. Neither Locke nor Martin has an extensive theory of motion, but Locke has such sensible things to say about it, even though what he said has turned out false. And there was no way Locke could have foreseen this.

I quote Locke illustrating his view, the illustration being somewhat clearer than the general statement that precedes it.

> The chess board, we also say, is in the same place it was, if it remain in the same part of the cabin, though perhaps the ship which it is in sail all the while; and the ship is said to be in the same place, supposing it kept the same distance with the parts of the neighbouring land, though perhaps the earth had turned round and so both the chess-men, and the board, and the ship have every one changed place, in respect of remoter bodies, which have kept the same distance one with another. But yet the distance from certain parts of the board being that which determines the place of the chess-men, and the distance from the fixed parts of the cabin being that which determined the place of the chess-board, and the distance from the fixed parts of the earth that by which we determined the place of the ship; these things may be said properly to be in the same place in those respects; though their distance from some other things, which in this matter we did not consider, being varied, they have undoubtedly changed place in that respect.
>
> (Locke [1700] 1979, Bk. II, ch. xiii, §10)

Thus, Locke tells us that *place* consists of (directed) distance relations among bodies and that motion is *change of place*. I want to split this into several theses and look carefully at the more obvious-seeming of them. There are two ways in which Locke's views run into familiar disputes: he takes motion and rest to be defined on bodies and commits himself to the symmetry of the relations of rest and motion. These themes are not my main concern. So I need to spell out this eminently sensible view in finer detail than is usual to focus on the target. Locke really puts forward a revision of an older understanding of what rest and motion are. This older understanding can be focussed in a definition and a thesis, as follows:

> *Topological definitions:* x is at rest if and only if x is in the same place at different times: x *moves* if and only if x is in different place at different times.

Directed distance thesis: x 's (relative) place is its (directed) distance from some place.

Locke then advances a familiar thesis which is highly plausible on both epistemological and ontological grounds:

Bodies thesis: the relations of place, rest and motion are to be defined on physical bodies. Any body, or group of bodies which maintain constant directed distance relations to one another, is acceptable as a place.

Its function is really to tell us *how to apply* the topological definitions. With the directed distance thesis, it gives us the theory Locke really wanted. I will call it the

Strong distance change thesis: If two bodies are changing the (directed) distance between them, then either is in motion relative to the other.

It is plausible and elegant but, of course, contentious. It implies the symmetry of motion among bodies, which may seem to be all that rescues it from being a platitude. For there is a weaker, hence more plausible, version of the distance change thesis which looks a thorough platitude. It is this:

Weak distance change thesis: If two bodies are changing the distance between them, then in every frame of reference, at least one of them changes place.

But I will argue that even this and, therefore, the directed distance thesis used to derive it are questionable and arguably false. When I mention the distance change thesis it is this weak version of it that I will mean, since it is so much more obvious-seeming than its stronger cousin. If it fails, we might feel that our whole ordinary conception of motion has been undermined (as was suggested to me in discussion of the paper). However, there is more assumed by Locke's account of motion than meets the eye.

Nevertheless, it is a simple account of motion. First, and by far most significantly, it is purely kinematical. There is a sense of relative motion in common use which is not merely weaker than Newton's absolute motion through space, but also weaker even than motion relative to an inertial frame. It is this that Locke wants to describe to us and, it is arguable, this that Newton meant by relative or apparent motion. It is the idea that we may take at rest whatever body we like. Then, without bothering to keep the causal books, let alone do serious physics, we can speak of things as moving. Further, we can do this globally, for any object anywhere, if we want to relate it to the one we chose. We *need* not ask about causes and forces. We can simply look for the vector changes that constitute change of place as we defined it. It is immensely plausible that this idea makes sense, and it deserves an account of its own because of its long life in our commonplace thinking. In fact, the very distinction between kinematics and dynamics just is the recognition of the naturalness of this idea of what motion is. But even when we weaken it by deleting its claim of symmetry, it is not beyond challenge.

Second, Locke makes no use of space as a continuant object, which is, somehow, to define places over time. He gives a neat *observable procedure for applying the topological theses* . Places are defined by bodies, so that any scheme for fixing places at different times has only to preserve the spatio-temporal continuity of bodies and the metric of the directed distances between them. Thirdly, the directed distance thesis and the weak distance change thesis look as certain as anything about rest and motion could look.

Doubts about the bodies thesis have been felt for a very long time. Notoriously, Newton rejected it on the basis of mechanics. Despite their rejection of Newton's version of absolute space and motion, classical physicists after Newton were never able to recapture the thesis for mechanics. Up to a point, it is quite clear why. Motion is certainly a causally significant feature of the physical world, but it does not have the *general* causal relevance which one would expect *a priori*. Though all motion is change, according to our topological definitions, not all motion is caused. It is too easy to overlook, in retrospect, the shock of Galileo's insight that inertial motion is a change without a cause. You must not look for the cause of it but rather for the causes of *change* in it. It is acceleration which is causally relevant.

Thus classical physics presents an awkward dilemma. An Aristotelian, absolute concept of motion in which motion is always a caused change (and so just like any other change) would have been elegant and plausible, but it raised intractable problems for an understanding of unaccelerated motion. A Lockean non-dynamical (purely kinematic) concept, giving rise to a strong form of relativity of motion, would have been elegant in a different way. But, of course, the salient fact about *forces* (causes) is that they *result in accelerations*; forced objects *had* to be regarded as accelerating. The upshot was a theory in which the criteria for rest (for identifying places across time) included a *causal* condition (the resultant of forces on rest objects is zero, as it is on uniformly linearly moving ones). This forbade our choosing any body we wish as defining the same place at different times, at least while we pay attention to causes. In fact, it might forbid us to choose *any body at all* as at rest, as Newton pointed out. This lends to space itself, an imperceptible entity devoid of mechanical properties, a key role in dynamics. Space fixes what bodies shall do when forces act upon them and what they shall do when force-free. It is obscure why attending to causes in the dynamic concept of motion provides a substantial role for an etherial entity, a role which the kinematical concept has no use for. It is an unsatisfactory, awkward state of affairs which resists an elegant explication.

The problems of the last two paragraphs are familiar enough. They spring from the bodies thesis. But they seem not to affect the plausible *distance change* thesis. The thesis says just that at least one thing must be changing place if one body changes its distance (or direction) from another. That seems obvious indeed. Yet, as I shall argue later, the development of Special Relativity raises a doubt about its certainty and General Relativity shows that it is purely contingent if true at all; furthermore SR does this *in the context of kinematics* where it is so plausible. Now it is clearly the *weak* distance change thesis which makes the step to the symmetry of Lockean motion seem so appealing, since the symmetry of distance change looks logically guaranteed. If x is changing its distance from y, then y must be changing its distance from x. At first glance, SR, through the Lorentz transformations, seems completely to endorse that symmetry in the symmetry of velocities among inertial frames. If F_1 is moving at v with respect to F_2, then F_2 moves at $-v$ with respect to F_1. But not even this is quite straightforward.

2 Presuppositions in the kinematical idea of motion

Despite its simplicity and obviousness, Locke's suggestion is properly a theory about rest and motion, which presupposes some constraints and extrapolates them. We can get some insight into this by focussing on a couple of obvious assumptions which, I think, any account of rest and motion must accept and then by contrasting Locke's theory with a not very plausible, but even more primitive one.

First we can and must assume that the concept of different places *at one time* needs no explication. It goes on puzzling me just what virtue of clarity lies in appealing beyond space to spatial relations. Our ability to count how many objects we see *at one time* is very elementary; it does not depend on a capacity to distinguish among them qualitatively. We simply see that they are touching or not, and if not, we see that there is space – a path or whatever – between them. We look directly across space from one to the other. On these topological relations, and the observation (at one time) of the spaces across which (and in virtue of which) they hold, all other spatial relations depend. So it is unclear how the relations might be conceivable independently of the space or vice versa. Usually, we count the things (at one time) by counting their several places. Locke actually obscures this fact (from himself as effectually as from us, perhaps) by analysing place in terms of distance. But this is wrong, I believe. Difference of place (at one time) is a *topological* idea, not a metrical one. We do not need to perceive metrical features in order to tell one place (at a time) from another. We see the different places directly. The role of metrical facts is different, as we shall see.

Next, every account of rest and motion accepts the following factual constraint. Places must be identified across time so as to preserve the general spatiotemporal continuity of persisting bodies or of the propagation of field quantities. (I ignore quantum problems.)

These then are two obvious presuppositions of Locke's proposal.

Let us now look at a theory, a very primitive one conceptually, of rest and motion in order to contrast it with Locke's. As far as I know, no one ever held this proposed theory. It is unsatisfactory, but contrasting it with Locke's will enable us to see that the directed distance thesis has a theoretical role and is not a mere platitude.

Consider this proposed account of motion:

A: *x* moves relative to *y* if and only if one or more parts of *x* each touch different parts of *y* at different times, or some part of *y* at one time and none at another.

The definition gives a basic account of such cases as an object sliding, rolling or walking over another. It is basic in a sense in which the distance-change thesis is not, because the definition presupposes nothing about a metric for space. Clearly enough, the definiens gives us a *sufficient* condition for motion, though we may need to remind ourselves that expansions and contractions must count as motions if that is to be so. But that is not unreasonable. If we accepted the definiens as a *necessary* condition of motion we would get a simple, but strange account of motion, since nothing would be in motion relative to y unless, at some time in the relevant periods, part of it touched part of y. But, of course, the account seems to settle, for every object in the universe, whether or not it is in motion relative to y.

Now one might argue that this is a perfectly coherent conception of motion even though it is, perhaps disconcertingly basic. What makes it coherent is the very same thing that makes it disconcerting. It is a topological definition of motion, whereas we ordinarily expect to get a metrical one. It certainly is a topological definition, since it defines rest and motion in terms of relations among bodies which are invariants of topology. The relation of physical contact does not exhaust the relations among objects which are invariants of topology. Enriching the set of relations considered would not solve our problem, however. The coherence of the definition is just the coherence of topology and our discomfort with it is just lack of familiarity.

But it does not work. For suppose x is sliding across y during the period t_0–t_1; suppose, too, that w is sliding across z during the same period. So x is moving relative to y and w is moving relative to z; but which of w or z is moving relative to y? Neither, according to the definition. A entails that both w and z are at rest relative to y, yet each in motion relative to the other. That is an unacceptable consequence of the proposed definition.

What is the source of this trouble? Not that place is a metrical concept. It is easy to distinguish places-at-a-time without a metric; they are given immediately. We can easily *define* the idea of change of place within topology: that is just different places at different times. But, if we confine ourselves to topology, we have too many arbitrary choices how to identify the immediately accessible topological places at one time with those at another, even counting each point of y as at the same place during the period. Counting y in the same place, and keeping all objects spatiotemporally continuous, we can still identify places through the period so as to count each point of w in the same place, or we can do it keeping each point of z at the same place. But there is no way to do it so that *both* w and z are at the same place throughout. Once we weaken A, then taking y as a rest body leaves the choice which of w and z to be at rest entirely arbitrary, as far as topology goes. So A gives no global decision as to what is at rest and what in motion, even after we choose y at rest. Taking y as at rest in a period leaves the problem that purely topological definitions can settle only *purely local* questions about rest and motion – questions local *just to y* – for there is no way, on the basis of topology, to extrapolate this further afield throughout the period. This, despite the immediate access we have to topological distinctions among places at one time. (We will find a quite similar problem for the distance-change thesis when we try to use it, in SR, except for inertial frames.) In particular, we cannot extrapolate the identification of places across time topologically just on a basis of continuity of bodies, even though any coherent idea of motion must presuppose that. The inadequacy of the idea is its inadequacy to provide for a global continuant space; its failure draws attention to a need for which Locke's provides. We need a way of showing *how to apply* the topological definitions which make up our oldest and most deeply entrenched understanding of motion. It is a quite similar kind of difficulty which besets the distance-change thesis in Special Relativity.

This makes clear what the role of the distance change thesis is in Locke's theory. We assign places and change of places metrically because nothing *less* than this general procedure will give us a foothold on how to assign places to things in space in a way which is even *topologically* coherent, and adequate to deal with rest and motion globally. The epistemology of Locke's theory, then, is not so simple, after all. The directed distance thesis and the distance-change thesis are not there because they are directly observable truths. Even on the face of it, these metrical judgements may be intricate and difficult. The theses meet a theoretical need which is very easy to overlook. A metrical account of place is required mainly because only a metrical account of *change* of place can give a coherent picture even of the topology of rest and motion. The epistemic gains and losses of distance measurements through the use of rigid rods etc. are of secondary importance.

This makes the relation of Locke's theory to that of classical dynamics considerably clearer; at least, it does to me. Both theories have a commitment to a global space in which every body is located and for which any scheme for identifying places at different times must be defined globally as well as locally. The difference between the two is that the link between force and acceleration requires that certain bodies not occupy the same place at different times. But if we take this seriously, we cannot guarantee that there is an acceptable way of identifying places across time which will yield *any* body at rest. Thus dynamics thrusts the commitment to a global space on our attention in a way which Locke's theory does not, though we have seen how the theory makes provision for a coherent topology for such a space.

I close this section with a brief overview of the history of our concepts of motion.

Before Galileo, motion was conceived as caused change of place in some form of container space. From this conception, Newton retained the idea that motion is change of place and he weakened the causal part of it: though uniform motion is change of place, it is uncaused. Later classical physicists weakened the idea of change of place instead, though they never quite abandoned it. One way to understand this is to see the attraction of holding that every kind of change is either caused, or else transformable away. So if one asks for the cause of the uniform motion of something, one could always invalidate that *particular* question by adopting a frame of reference in which the thing is at rest and not changing. That neither Newton nor later classical physicists abandoned the idea of rest as maintaining the same place in space is quite clear; an acceptable frame of reference is not required to possess a rest body. Thus the topological definitions are retained as our primary understanding of rest and motion. And, in fact, very few acceptable frames are identifiable by means of central rest bodies, despite persistent claims by empiricist-minded writers, especially in the early days of SR and GR, that inertial frames *are* systems of bodies. The history of our theories of motion displays clearly a frequent refusal among empiricist-minded thinkers actually to *look at the evidence* in order to see what the facts about rest frames in existing physical theory really are and an obdurate clinging to their own *a priori* convictions about how physical theories must construct their concepts despite the evidence of how they actually do so.

3 Kinematics in Special Relativity

Locke's apparently modest weak distance-change thesis rests on assumptions. We can see this if we look at some results in SR.

Special Relativity is special in that it restricts the range of reference
frames relative to which we may give physical (i.e. mechanical and
electrodynamic) explanations of what happens. The permitted reference frames are just those allowed by classical physics and called there
inertial frames of reference. The theory recognises absolute acceleration just as Newton did; but there is an absolute uniform motion, too,
in the special case of light (*in vacuo*). It moves, and moves at the same
speed, with respect to every frame of reference. From a different perspective, SR is just physics when spacetime has no curvature in it. It is
flat spacetime, rather than the use of inertial frames, that makes it
special. So let us see whether we can base kinematics on the bodies
and distance-change theses in flat spacetime.

In fact, we cannot base even kinematics on a concept of motion
which conjoins the bodies and distance-change theses in flat spacetime. More explicitly, the distance-change thesis is tied to inertial
frames, and the bodies thesis is possible only if one gives up the distance-change thesis, or, more accurately, gives it up as a thesis which
appeals to natural measures in a non-inertial frame.

That the *concept* of motion in SR is not the *concept* of distance-
change, even for kinematics, is clear from the addition of velocities
theorem. Given a suitable frame of reference F, suppose that two
objects a and b are in uniform motion in opposite directions from
some point in the frame. For convenience, let their speeds be high:
$\frac{3}{4} c$ for instance. Now we can ask *two distinct questions*, given this
information, one about distance change and the other about motion.
That the questions differ is shown clearly by a difference in the
answers and in the styles of calculating them. We can ask, first, 'What
is the rate at which a and b are changing the distance between them?'
The question will have an answer relative to the frame F, with respect
to which we gave the information. The answer is just the vector difference in each velocity relative to F. In our simple case, it is the sum of
the two speeds. So correct answers to distance-change questions can
give rates greater than c, the velocity of light relative to any frame. If
we now ask, 'What is the rate at which a is moving relative to b?' we

must not suppose that its answer is still to be related to the frame *F*. We have asked a question which refers us to a new frame, in which *b* is at rest and *a* is in motion. The information given allows us to calculate an answer, but it is not the simple calculation of vector difference we just used. We must subject *a*'s speed relative to *F* to a Lorentz transformation before we can add it to *F*'s speed relative to *b*. The resulting sum gives a speed always less than *c*. The use of high speeds in the example merely makes emphatic a general result that the relative speed of one to another is always less than the rate of distance change relative to a third system.

The two kinds of distance change in SR are not merely new-fangled versions of the classical distinction between real (or dynamic) motion and kinematical motion. The advantages of inertial reference frames are no longer just that they let us assign causes and forces more neatly, nor even that they yield linear paths for all vacuum light signals. The rate of distance change between two things relative to some frame fails to give us the rate of distance change of the one *relative to the frame of the other*. In taking the case of *a* and *b* in uniform motion relative to *F*, we considered only the simplest possible case, omitting acceleration. In the light of all this, how can one assume, in general, that the motion of either has *any* proper *metrical* description with respect to the other? What makes kinematic descriptions in classical terms clearly admissible, quite generally once we ignore causes, is precisely the simple metric transformations which we see fail in SR. Of course, it does not follow from their failure that *no* proper metric description *is* available. That needs more explicit argument aimed at putative non-inertial frames for SR. But the clear path from distance change to motion becomes obscured by this rather elementary consequence of SR.

SR is a theory in which we may not properly *apply* the distance change thesis to yield a relative speed *until* we have first selected a frame of reference – which means that we have already adopted a scheme for identifying places across time; that is, we have *already applied the topological definitions*. I read Locke's assumption as precisely the opposite of this; that we should not use the topological definitions of rest and motion in a principled way until we have used the bodies and distance-change theses to identify and distinguish places-at-a-time and change of place at different times. It turns out not to be a trivial assumption, for it is false in SR. In practice, perhaps, we choose frames by deciding which bodies are caused to accelerate, or by choosing frames relative to which all vaccuum light paths are linear. But the heart of the matter is that only inertial reference frames allow a satisfactory kinematics in flat spacetime. The strong distance-change thesis is false for kinematics.

Let us take the simplest case which poses the well-known and much discussed Clock or Twins Paradox of relativity theory. We suppose twin *R* at rest throughout in an inertial frame. Twin *M* passes him at great speed, moving uniformly. As they pass we note that *R* and *M* each has a clock which is, at that moment, synchronised with the other. After some suitable period of travel, *M* reverses his motion, then moves uniformly back to *R* at the same speed as before, (relative to *R*). As they pass we note that *R*'s clock reads later than *M*'s does. I take it that this brief exposition is enough to make clear what is a familiar example.

We are apt to have strong intuitions about kinematical symmetries between *R* and *M*. We naturally think that whenever *M* is in uniform (accelerated) motion relative to *R*, then *R* is in uniform (accelerated) motion relative to *M*. However this sense of the obvious is mistaken; the apparently obvious claim is not even a well-formed sentence of SR. For our 'whenever' is a universal quantification over times, but does not say whether the times are the simultaneity classes of *R* or of *M* or some other frame. In fact, when we get some well-formed claims to replace the single symmetry claims – and we will need two in order to speak of *R* times as well as *M* times – the resulting sentences are in fact *not* both true (see Nerlich 1994, pp. 260–3).

That the Clock Paradox is about kinematics is clear from a simple argument that accelerating forces are irrelevant to the discrepancy between *R*'s and *M*'s clocks. The accelerating and decelerating forces to which *M* and his clock are subject, occur between periods of his uniform motion relative to *R* , during which he suffers no forces. So these forces imposed on *M* will be just the same whether he accelerates and decelerates in this way on a trip to Jupiter or on a trip to Sirius. Only the periods of uniform motion will differ. But the slowing of his clock is a function of the total length of his trip, not the way he accelerates and decelerates. So there have to be differences in the *kinematic* description of the episode with respect to *R* and to *M*.

This does not yet show us that *M* cannot retain the distance thesis about *R*'s kinematic motion in a natural way. But it ought to erode our confidence that taking M at rest gives us a quite simple transformation from R's kinematical description of the episode to one that is open to M.

Indeed, M will *not* be able to give a consistent kinematical account of R's apparent motion if he tries to do it in the most obvious way. What, first of all, is the most obvious way? Well, it is for M to identify his frame of reference with the inertial frame with which he coincides during his linear motion on the outward leg – that is, he naturally identifies himself with that system *for the period in which he coincides with it*. Later, he coincides on the homeward leg with a quite different inertial frame with which he will naturally identify himself – naturally, in each case, because he will be able to measure what goes on round him with his own rods and clocks in a customary way in both cases. M's problem is to make consistent, combined, global sense of what he measures.

But no such combination will work. There is a quite general kinematical difficulty which makes the proposal absurd. No two inertial frames can combine their measures globally. The hyperplanes of simultaneity of the two systems will intersect somewhere; they will meet in a subspace of events. (In a familiar two-dimensional space-time diagram, the lines of simultaneity meet in a point, or in a line for a three-dimensional one.) For any event, E, in this subspace, there will be simultaneous distant events, E' (by a simultaneity inherited from the first frame) and E'' (by a simultaneity inherited from the second frame), such that E' is absolutely earlier than E''. So pairs of absolutely non-simultaneous events will nevertheless be simultaneous with the same event, relative to the putative combined frame. This is absurd. That is to say, generally, we cannot integrate two frames of SR together into a non-linear frame and escape contradiction. The contradiction is kinematic.

However, it is important to understand just what the example shows. It does not show finally that we cannot take M as our frame of reference, but that we can do so only at a price. The wish to see M as a possible frame is the wish to preserve the bodies thesis. But the price of this is that we lose the distance thesis, at least as *something epistemically accessible*. To make this clear, we need to look briefly at the sort of options open for M as our frame of reference.

We struck the inconsistency just noted (and are open to some others not yet mentioned) because we tried to retain a *rigid* system – so called, because it is a system in which, first, rest points keep a constant distance from one another as measured by the rods of the moving system, and second, the time interval between classes of simultaneous events is the same at every spatial point as measured by clocks of the system. Our problem is that we cannot combine rigidity and curvilinearity in the geometry of Minkowski spacetime. That is due to the global symmetries of its flatness everywhere. So rigidity has to be sacrificed once we try to regard *M* as at rest.

To see what that means, consider one manoeuvre open to us. We can give linear time-like ordinates and flat coordinate hypersurfaces, rewriting physical laws in covariant form as in GR for use in the corresponding non-inertial frame. We can 'contain' *M*'s curvature locally by taking nearby time-like ordinate curves (representing rest points of the nonlinear frame) progressively straighter, so as to localise the curvature of *M*'s worldline. Most regions of the coordinate system will be Lorentz and the corresponding frame largely inertial. But we will also have to prevent our simultaneity spaces (hyperplanes) from intersecting by letting them curve, locally, too. But now we will find, in our non-inertial frame, that there are pairs, *p* and *q*, of spatial points, each deemed rest in the frame, yet such that the distance between *p* and *q* is different at different times. That is the direct result of the local curving of coordinates; it sacrifices rigidity in the frame. Each *p* and *q* constitutes the same place at these different times. They are *rest* points. So the distance between them changes, *without there being any motion*, according to our frame description. Thus, not just the strong, but also *the weak distance-change theory would be falsified*, or, better perhaps, *forsaken* so that we can keep *M* at rest. Whatever we call it, there can be no question that this option is not consistent with taking the weak thesis as an *a priori* or necessary truth about motion. It is worth noticing (in passing) that our frame will also oblige us to say that time runs at a different rate at *p* from the rate it runs at *q* (or at some other points, if not these two). Yet this attempt to cling to the bodies thesis has to reject the distance thesis: things in the deviant frame change the distance between them, though neither moves; and, conversely, of course, things can move yet not change the distance between them.

It might seem that there is a natural alternative to this unwelcome outcome of allowing curved time-like and space-like lines (hypersurfaces) to define same place and same time for the frame attached to *M*. Given one 'accelerating' point (twin *M*) which we want to take at rest, let us select the others so that they are always at a constant distance from the first, *as measured by the end points of an unstressed rod* that co-moves with *M*. This gives a different clear sense in which we hope to retain, for *M*, a rigid frame of reference. The consequences of this alternative are quite well known (see Rindler 1977, chapter 2.16). They present us with an intractable objection. In proper acceleration, a point has the same acceleration relative to each inertial frame at the instant at which it is (momentarily) at rest in the frame. A set of such points move together like an unstressed rod. Can this make the basis for a kinematical frame, usable globally, in flat spacetime? No. What works locally does not work globally. There will be a critical point, *O*, in the putative frame with which we would like the rod to co-move. We cannot suppose that the rod (or any other material thing) contains that point, for then this material point *O* would have to be a photon and move along a lightlike trajectory. Points beyond *O*, which retain a constant rod-measured distance from *M* would trace out spacelike trajectories. Not only could these not be the points in any material rod at this place in the putative frame, they could not possibly be regarded as defining the same *unoccupied* place at different times, either. So the frame could give us rigid measures only quite locally. This still holds if the proper accelerations of the set of points vary together in time. The result is an absurdity. This alternative is not possible.

Other shifts might be resorted to, but all of them have some such objectionable features. Let me stress again that the argument shows that we cannot describe motion *kinematically* relative to accelerated frames in flat spacetime. The discussion in this section mentioned forces, masses and the concepts of dynamics only in the paragraph where I argued their irrelevance to the Twins Paradox. In Newton's terminology, it is not just the facts of *real* motion in SR which are inconsistent with body and distance-change theses: apparent motion is inconsistent with them, too. What works for local kinematics does not work globally.

The result of this discussion is that for SR (that is, for flat spacetime) the bodies thesis is incompatible with the strong-distance thesis. If two bodies are changing the distance between them, then, even for apparent motion, it may be impossible to take either at rest. For if neither is at rest in an inertial frame, we cannot sensibly describe their kinematic motion. Of course, each will be in motion relative to some inertial frame. At least, that is so in a simple physics in which distance change is tied to simple and natural ways of testing it.

How does this differ from classical physics set in Newtonian (or post-Newtonian) flat spacetime and why? In classical physics inertial frames are privileged by criteria that belong within dynamics: a frame is inertial if all accelerations relative to it can be tied to sources in a meaningful and confirmable way. In short, it is inertial if the three laws of motion come out true with respect to it, and with forces tied to material sources in simple ways. The frames are indistinguishable from one another not because there is no meaningful structure which, in principle, could possibly distinguish them, but because the geometry of space is Euclidean and strongly symmetrical. But consider the standpoint of the purest form of classical kinematics, where all that concerns us is to plot the metrically defined motions (distance changes) of some body or bodies from the standpoint of a randomly chosen frame of reference. The kinematical relativity of motion is obviously perfectly general. There are no frames privileged for this purpose. This depends, still, on the Euclidicity of classical space, but we have cut the link with force and motion. I take it that this fact has very much predisposed philosophical thinking toward the view that there must somehow be a general relativity of motion for mechanics in general.

But, of course, the general relativity of pure kinematical motion in classical physics also depends on absolute simultaneity. In classical theory, it is clear that at any time when a is in uniform (non-uniform) motion relative to b then b is in uniform (non-uniform) motion relative to a. The failure of that symmetry in SR is the most obvious disanalogy with the classical case. Now it might seem that, in the transition from classical to SR physics, what happens is that the structure of absolute simultaneity is *deleted* so that Minkowski spacetime differs from classical ones in lacking the fibration of spaces at times in which all classical spacetimes are alike. So there is less structure in SR than in Newton's physics.

But that is not the case. Even Newton's spacetime with absolute space is defined only up to an affinity (see Nerlich 1994, §10.4 for an intuitive account of this). It is the *lack* of structure there which allows the partitioning into spacelike sheets and permits us to adopt curvilinear frames without absurdity. Minkowski's spacetime is a metric spacetime, fusing spatial and temporal measures together. It imposes global pseudo-Euclidean symmetries across the whole structure. These distinguish the linear trajectories as uniquely suitable for global reference frames. The relativity of motion for kinematics fails in SR because it is a factually richer structure: richer, specifically, in its geometry, its spacetime structure. It is the comparison between the two kinds of whole spacetime structure which illuminates the differences here and *explains why* the relativity of motion is so much more limited in Minkowski spacetime. A comparison among the objects and their distance and angle relations can shed no such light unless we deduce a spacetime structure from them and exploit its geometrical features to illuminate motion.

Does all this drop into a quite new perspective in GR? The main new element is this: in flat spacetime (as we saw) there *are* frames of reference which retain rigidity in an intuitive way, even though they do not allow us to keep the bodies thesis. They are linear or inertial frames. So we regard those based on curved (accelerated) time-like lines as unacceptable. And quite rightly. But if spacetime *itself* is curved, we easily see that it may be impossible to retain the distance thesis, whatever we do, since *all* timelike geodesics may diverge or converge and *none* be parallel. Spacelike hypersurfaces may curve so that the time intervals between them are different at some places from what they are at others, however we choose coordinate systems (except for choices sharply confined to what is local). The space *itself* will expand or contract, not merely have curved coordinate lines and surfaces through it. So a GR spacetime may make all global reference frames non-rigid, because it lacks the symmetries of flat spacetime. Rest points will change the distance between them not merely because we chose a perverse frame in order to cling to the bodies thesis, but because there is no rigid choice open other than locally. Thus, GR is closer to the *bodies* thesis than SR because it must envisage abandoning the *distance change* thesis right from the start. So even the weak distance change thesis is false in the general run of GR spacetimes, even when we free it from association with the bodies thesis. Certainly, GR offers no aid whatever in retaining their *conjunction*. Further motion is not generally a symmetrical relation in GR, so the strong thesis is falsified too.

4 Some epistemological conclusions

I am suggesting that the mistake in Locke's analysis of motion is a mistake about kinematics, which means that it is a mistake about the distance-change thesis rather than the bodies thesis. He was wrong just where it looks as if his ground is safest. Now, just what *kind* of mistake is that? Clearly, whatever kind it is, Locke had no chance of detecting it, and I do not mean to patronise him (or anyone else) in speaking as I do. Indeed, it is an *empirical* discovery that we can never be in a position to make the kind of epistemological judgements which Locke himself felt thoroughly entitled to make.

Let me contrast Locke's error with the older mistake in dynamics, that motion always has a cause. That error supposes that an epistemically immediate concept of motion, given by the bodies and distance-change theses, always has a cause. It assumes that all motion must be caused because all motion is *change* – change of place. If we take it that Locke's analysis of kinematics makes a false assumption about the way of the world, then at least it could not be a causal assumption, but lies deeper and more hidden. So it is not because Locke underestimated the causal complexity of the world that he fixed on the distance change thesis, but because he overlooked something more primitive than that, something on which we need a still sharper focus.

What Locke could not have foreseen, then, is that, even for flat spacetime and for kinematics, we cannot use the distance change theses until we have applied the topological definitions. If we do not apply them first, our local distance measures will extrapolate either absurdly or arbitrarily, just as happened with the absurd or arbitrary extrapolations of the primitive topological definition of §5.1. In that primitive definition, decisions on rest and motion are well defined locally by contact with the central rest body and absurd or arbitrary beyond it; in a non-inertial frame, local distance measures apply where coordinates do not intersect, etc., but are absurd or arbitrary beyond that region. So the distance change thesis is globally usable only in inertial frames. Variably curved spacetimes differ from flat ones because there may be *no* way to extrapolate local distance measures rigidly, because space itself expands and contracts. We are comparatively free, in GR, to take bodies as at rest, but this is because the distance change thesis is *false*, not because it is true. If space is expanding then the distance between rest points changes not because of a quirky choice of reference frame but because of the structure of the thing itself, spacetime. No frame has the advantages of global rigidity.

One reason why it looks particularly hard to foresee all this, even in the case of flat spacetime, is that every means for applying the topological definitions presupposes at least local distance measurements. Criteria for an acceptable SR frame – either the linearity of all vacuum light paths or the causal principles involved in the first law of motion – each require local test by the use of metrical techniques. From a strictly conceptual standpoint, the very general *projective* structure of spacetime guarantees that the spacetime trajectories of vacuum light paths and of force-free particle are geodesics, but I cannot see how this offers a *practical* way of applying the criterion which does not require local metrical presuppositions. I have in mind complications of the sort discussed in §4.2 and §4.5. The technique of clock and light-particle interactions for measuring spacetime intervals is given a simple treatment in Geroch (1978), chapter 5, esp. pp. 80–96; and is very briefly outlined in §3.5.

The distance-change thesis makes *local* sense of kinematics, however we apply the topological definitions and whatever the global structure of spacetime (provided that there is no local singularity). That is, it makes local sense globally, so to speak; you can apply the body and distance-change theses anywhere, anytime and get a coherent local kinematics. It is easy to make the following mistake about SR: Its kinematics is the conjunction of the bodies and distance-change theses, just as it is in classical physics. It involves an assumption, which easily eludes notice, about how to extrapolate what seems to be an analytic claim about rest and motion, based on a definition which *is* adequate locally. However, there are no natural global length measures in one's frame till one has shown it to be inertial. The naive extrapolation – the one that seeks to bypass prior application of the topological thesis – breaks down despite the isotropic structure of flat spacetime and despite the universal propriety of using local distance measurements anywhere, anytime and in the same way, to discover acceptable frames of reference. In fact, the extrapolation fails rather *because* the global structural symmetries of flat spacetime provide so simply for global projection from the special basis of inertial frames. What distinguishes these frames in SR is, properly, that they provide this general metrical advantage. That is why Locke was wrong about motion.

Two more general epistemological morals can be drawn from this,

I think. The first is that the ways in which apparently necessary truths can turn out contingent and even false are too various for epistemologists to hope to foresee. Contrast the present case with the Causal Principle or with the status of Euclidean geometry, or even with the closely linked failure of the concept of absolute simultaneity. To begin with, each of these appears as a *thesis*, albeit *a priori*, not as a definition. Each looks as if it has a truth-maker, whereas the distance thesis looks like an entirely feasible *stipulation*. The Causal Principle seems to have been the shadow in the material mode of a requirement in the mode of thought that each event should have an explanation. The second and third seem to have been failures of imagination in which the (apparently) familiar and the simple are mistaken for what *must* be so. But in our case, both the bodies thesis and the metric extrapolation from the regional to the global in SR seem positively invited by spacetime homogeneity. The prior loop through the topological definitions has to look like an eminently dispensable epicycle. Things appear almost to conspire in masking the truth even once one has some grasp of SR. Other assumptions of ours might turn out delusive in similar or perhaps in quite different ways. I see no hope of certainty that one can define concepts in ways that seem epistemologically advantageous without all manner of existence and extrapolation assumptions.

Lastly I conclude that, in our theories, *understanding* is considerably more important than perceptual immediacy. We do understand the topological definitions quite readily, though we cannot apply them across time on the basis of simple observation, and their role in SR is easily grasped. But what constitutes *understanding*? Understanding shows in the *capacity to generate explanations*, examples, speculation, and in the capacity to see what the limiting cases are and which are the crucial observations. This is a vague and consequential account because understanding is not given in any set of algorithms. We cannot cash understanding in terms of familiar shibboleths such as deductive power in a formal system and the like. That sort of cashing smacks too much of the philosophical motives that lead us to prefer perceptual immediacy and the sort of hard-headed empiricism that goes with it. The trouble with such hard heads is that they may contain rigid minds fixated on various a priori epistemic ideals. It is just such motives that I am concerned to cast doubt on and to demote below understanding. I should prefer to link understanding to the power to visualise, taking the ideas of space and spacetime as arenas which give us precisely that power. It is, I think, the sheerest prejudice and lack of imagination to suppose that space, because imperceptible, cannot be an aid to visualising. One visualises motion and rest in space with great ease and generative power simply because one treats space as having many of the properties of bodies – extension, rigidity, continuity for example. The capacity to visualise is, no doubt, not the only source of understanding but it is a main one and deserves more considered attention.

Part Two

Variable curvature and General Relativity

The essays of this part form a somewhat less unified group than those in Part 1. In some ways the ideas are more sophisticated because the geometries they deal with are no longer Euclidean or pseudo-Euclidean. Yet the first two papers are among the most generally accessible of those I have written on space and spacetime. Only the easiest of departures from Euclidicity are considered, they concern only space and they are developed in quite intuitive ways.

A powerful force which does much to shape the literature on space is the abhorrence felt toward it because it is unobservable. We are much inclined to think that this characterises its essence, its ontic type. That is, if there is such a thing as space (or spacetime), then it is of a kind deeply different from the kind of thing generally admitted as real in physics. The deep difference is precisely that it is imperceptible in some principled way that springs from its ontic type. A main theme in §§6 and 7 is that this is a deep error. It is because of its geometric type that our concrete, actual space eludes perception. It's because it is, locally at least, very nearly Euclidean.

It was confusion over this which led Leibniz to argue that we can neatly detach the spatial relations which things have to one another from the relations which they have to space. Robbed of its perceptible, material framework, space stands naked before us, its ontic type shockingly revealed: it is a bogus entity, not a real thing at all. But Leibniz's argument fails. This is plain once we apply it to a thing of a different geometric type (even though of the same ontic type). In spaces where the curvature is not zero, things are observably dependent on their relations to space for their observable spatial relations to one another. That spatial symmetries conserve material structures is apparent at once. The problem lies deeper than the falsity of Leibniz's premise, the ground on which his argument is most often criticised. The argument itself is invalid. It fails whatever the geometry, since the principle of the argument, the Identity of Indiscernibles, is plainly false.

This raises a difficulty for relationism as a reductive philosophy. The theory substitutes for space just objects (or perhaps occupied spacetime points) and spatial relations among them. This depends on the view that we may speak of spatial relations among things without begging the question whether spatial relations are of a kind such that they can hold without a space to sustain them. On the face of it, it is highly plausible that spatial relations do depend on space. Intuitively, they are mediated relations – if one thing is spatially related to another then there is a path between them. A path, whether materially occupied or not, is a part of space and of the same metaphysical type as it. We feel the pull of the very same intuition, but from a somewhat different angle, when we find it plausible that if one thing is at a distance from another then there is somewhere between them, whether or not anything is at the place. It is Leibniz's detachment argument which has traditionally been supposed to extricate spatial relations from their dependence on space. I argue that it does not succeed in this task. Relationism begs the question against realism.

By taking a more or less classical world and changing merely its geometry in one or two simple ways, one can see that the role of geometry in explaining states of things, events and processes might be much more prominent than it actually is is. This also enables us to see that, even as things are, it explains a good deal. It accounts for more than we naively suppose. Given that our space is close to Euclid's, its symmetries conserve features rather than changing them – for instance, things don't change their shape when they change their size and their relative motions make no differences among them. It detracts nothing from the significance of this for spacetime explanations that the examples have a mainly classical setting: the examples can be set in a relativistic context without change, but the classical setting is more widely accessible.

It is important to see that spacetime exploits a style of explanation which is not causal, or, if you prefer, not causal in quite the usual sense. There is something distinctive about it and I try to show what that is. A problem here is that we have no real consensus about what counts as causal explanation and what does not. In the context of some specific theory or other we can see clearly enough, I believe, what the peculiar explanatory role of space or spacetime is. For instance, in theories of mechanics, causes are forces. Given the right examples it is easy to see that some events or processes which can be quite clearly explained strictly within the specific theory of mechanics in question are nevertheless not explained by forces. This may either be because there are no forces at work anywhere or because the way the forces are involved in the process is itself to be explained by the spatial structures which define the directions and distances across which they act.

Comparing these worlds with Newton's reveals just what is unsatisfactory about his world. It is not that it involves obscure or ghostly entities which cannot play a proper role in the physical world because they are of the wrong ontic type to play it. Newton's world is simply unsatisfactorily bland. Its space is Euclidean. This may give us an uncompelling motive to look for an account of the world which does not commit us to these symmetries which make so much perceptually elusive. But, in the classical context of space and time, it is hard to see a truly clear, intelligible alternative to Newton. The strength of his view of things emerges plainly in the contrasts drawn in §§6-7. When we look at Newton's world as a spacetime world, then of course the picture is quite different. What we see then as unsatisfactory about it is simply this: it is awkwardly indefinite. We expect spacetime to have a metric, but classical spacetimes have only affine structure. Here again, this does not mean that the world couldn't be definite only up to affine structure. There is no logical or ontological objection to that. But it does make the world unsatisfactory, and metaphysicians unfortunate enough to dwell in such a world would be doomed to interminable, indecisive bickering over rival ontologies. But if the relativity theories are correct, we live in a satisfactory world in which we can see that the truth lies with realism.

There are other ways in which those theorists for whom the medicine of realism is too strong have sought to limit its claims. I speak of the conventionalists who, as I see it, were so much, but so wrongly influenced by Einstein's view that all attempts to settle simultaneity in a frame of reference are trapped in circular procedures. The only exit from them is by means of some arbitrary definition of simultaneity. It was this thought, rather more than any other, which gave rise to the view that the advent of relativity was the triumph of Ockhamism. In §8 I try to show how mistaken this view is and how much it impoverishes the content of physics. Not only does this cast the metric of space as a mere convention imposed on spacetime for obscure reasons. Not even affine or projective structures are well defined. In particular, I consider the idea, so crucial for the factual import of GR, that the curvature of spacetime is a real structure and I try to trace just what its spacetime foundations are.

Leibniz's detachment argument still plays a role in metaphysics even in the deeper and more sophisticated realms which are at issue in tracing the structure of GR. The Hole Argument is a novel application of it in contexts which bear no very obvious analogy to those in which Leibniz first used it. The Hole Argument seeks to do away with structure but in ways quite different from those of the conventionalists. It is remarkable, in fact, for aiming to keep the upper levels of spacetime structure but for ridding us of the most basic one of all, the manifold structure, the smooth topology of spacetime.

The context is certainly much more complex and more theory laden than in anything Leibniz ever wrote to Clarke. But the fallacy, I argue, is essentially the same. It fails to make sense of the explanatory role of symmetry. In the context of Leibniz's argument, the blandness of Euclidean geometry is not at all the sign of a bogus, a dispensable or a non-existent entity. This is plain once we substitute for it an object which lacks the symmetries. Exactly the same is true of the bland manifold. Its non-trivial conservative role emerges in exactly the same way: imagine an asymmetric manifold and the conservative role of the standardly assumed symmetries is at once plain and plainly physical. The more sophisticated setting of the Hole Argument fails to protect it from the fallacy Leibniz made in the seventeenth century.

6 How Euclidean geometry has misled metaphysics

1 Introduction[1]

Euclid's geometry is so brilliant that it has always dazzled us. It shone for centuries so brightly as to blind us to any model for an established branch of knowledge but the deductive system based on a few self-evident axioms. Thus it misled metaphysics. But that is not my theme. I want to show how it led Leibniz to some mistaken arguments in the central letters of the Leibniz–Clarke Correspondence. The letters are about quite special issues in the metaphysics of space. But they have dominated philosophy since then as our models of how epistemological arguments should decide metaphysical questions. My theme is to reveal the arguments as mistakes.

Space sets ontology one of its most acute, searching and elegant problems. To change my visual metaphor, the reason why we have not been able to see further into this subject is because a giant has been standing on our shoulders. He is Leibniz. Though the arguments in his letters are fallacious, surely he is a giant. What can make those fal-

1 I have read versions of this paper to a number of meetings in Australia, Canada and the United States. It owes much to the stimulating discussions it received. I am especially grateful to my colleagues at UC San Diego where I wrote the first version of it while on study leave there.

lacies clear to us was not available to him nor to his great opponent, Newton. But first, what is the problem space sets for ontology?

On the face of it, space has disconcerting properties. The list below is loosely phrased, and neither exclusive nor exhaustive. But it shows us what the problem is – on the face of things, at least:

(*a*) Space is a concrete *particular*, being neither general nor abstract. It has parts, not instances. There is only one space (we ordinarily think) within which everything is related to every-thing else.

(*b*) Space stands in *concrete* relations to concrete things. I *move through* it, *see across* it to trees and cars at various distances, it *contains* me and all these concrete things. These are relations which, typically, only concrete things have to each other. Abstract things can play no part in them.

(*c*) Space is *physical* – that is, space plays a prominent role in physics. Physicists are always tracing paths and trajectories, mea-suring lengths or constructing theoretically sophisticated, elab-orate and expensive standards of length invariance. It also fig-ures in their theories.

(*d*) Space is *not material.* As classically conceived, it has no mass, momentum, or energy. Our usual nomenclature for views of the nature of space is unhelpful. Newton denied that space is a substance just because it is not material in this sense – 'it does not support characterisitic affections of that sort that denomi-nate substance, namely actions such as are . . . motions in a body.' That seems right. If so, then the view that it acts, is mate-rial, or has mass or energy, surely ought to be called substanti-valism. That is the right way to describe spacetime as we find it in General Relativity.

(*e*) Space is *not perceptible.* We never see space or its parts; we nei-ther hear them, smell them nor touch them. Yet we can see across it. We can see at a glance that just two objects are now visible though we 'can't tell them apart'. Suppose that they are in fact qualitatively identical. That surely means that we can see immediately that the two are in different places even though we see only the things and not the places.

The acute problem for ontology lies in the sharp clash between the first three properties on the list and the last two. The first three properties seem to tell us that space is something; the last two that it is nothing. This problem about space in our ontology is unresolved. Realism is my name for the view that space is a thing, earns a place in our ontology and has (something like) properties mentioned in the list. I distinguish realism from *substantivalism*. That is the view that space is a kind of real thing, one which acts on or is acted upon by things, and that it is a continuant.

The precision and power of geometry make the space problem elegant because the arguing can be so neat, clear and determinate. The curious relations of space to perception (feature (*e*)) demand that we frame with particular care and rigour the epistemological principles on which we either reject or accept an ontic commitment to space. So the problem is searching too.

2 Doubling in size

Let us begin with a familiar and graphic challenge. It snugly fits the mould of Leibniz's arguments, though we owe the first form of it rather to Poincaré. Both the style and the mistake are the same as in Leibniz's letters. Every philosophy student surely hears this argument at an early stage of his education. It is one of our best-known intuition pumps. It will be useful to start with it.

Suppose everything were to double in size overnight. Nothing would betray the change. Round pegs would fit snugly into their old round holes and other holes squarely resist their entry as before. Everything would be related just as it was to everything else. The difference would really be no difference, the supposed change no change at all. Nocturnal expansion is absurd.

In more detail, the argument has to run like this. At first glance space looks a useful posit: the size of a thing is the quantity of space it fills, its shape is the disposition of the thing's parts in space, the spatial relations of one thing to others is fixed by their places in space and the lengths and directions of the spatial paths that join one place to others. The spatial relations between things look as if they are all mediated by relations of things to space.

The doubling argument seeks to cut this tie. It is a kind of reductio argument in which the absurdity is not a contradiction but an idleness. It is a thought experiment in which the realist gets his space but finds that he can do nothing with it. It shows how to change the thing–space spatial relations of everything yet retain all thing–thing spatial relations. Conversely, the thing–thing spatial relations are shown not to determine any thing–space spatial relations. The relationist's gift of space is revealed as ironic – it gives nothing at all. There are no thing–space relations. So it is needless to posit an entity that exists only to found them. Space is a chimera.

All this rests on some kind of indiscernibility or inconsequentiality principle. It might be this one:

IP: 'Inconsequential properties' are not properties

Here 'inconsequential property' means something like 'property which can change without any accompanying change in specifiably privileged properties'. For the case in hand, we have two ways of saying what is privileged. Observable properties are privileged; more specifically (and less naively) thing–thing spatial relations are privileged. I will stick to the latter account, which will do for Leibniz himself. So let us distinguish thing–thing spatial relations from thing–space spatial relations. Thus we can state the detachment thesis:

Detachment thesis: thing–thing spatial relations are logically independent of thing–space spatial relations.

This will put us in a position to deny the realism of the introduction. All thing–space spatial relations are idle; so the relatum of these relations is idle too. Space is no part of our ontology. Only the concrete things that stand in spatial relations together with those relations themselves are real; space is an unwanted absurdity.

The detachment thesis lies at the foundations of a prominent reductive theory of metaphysics: spatial relationism. Spatial relationism can fairly promise to reduce space to thing–thing spatial relations as long as these relations can be detached from thing–space spatial relations. The detachment thesis clears relationism of the otherwise plausible charge that it is circular. The detachment thesis is no help in showing how the relationist can complete his programme. It simply legitimates his title to use thing–thing spatial relations in the reduction. It offers an escape from circularity. It is not obvious how to escape circularity without it.

3 Objections to the doubling argument

Perhaps the reason why Leibniz never used the doubling argument is that he read the first of Galileo's *Two New Sciences*. Galileo points out that doubling a thing in all its lengths (scaling it up) will quadruple its cross-section areas and multiply its volume and thus its mass by eight. The load-bearing factors of the object are essentially functions of area, but the loads they must bear for the thing to be stable are functions of its mass. So scaled up models of strong small things, like ants, may be unstable, as producers of gigantism horror movies often find. The bigger they are the harder they stand – or fall. What, too, about the speed of light, and other constants – Planck's for one?

Poincaré thought that he could avoid this objection, and it is evident that Reichenbach, Grünbaum (1967) and many others thought so too. This was challenged by George Schlesinger (1967). I will not try to umpire that debate, though it surfaces again briefly in §6.7, objection 2. My main objection, in the next paragraph, is simpler and independent of this. I think it is more incisive.

The claim that doubling has no consequences rests on a quite precise and unique assumption about thing–space relations. Things must be doubled *in Euclidean space*. Among the geometries of constant curvature, only Euclid's geometry allows similar figures of different sizes. If everything doubled in size overnight in a non-Euclidean space everything would change in shape. Thing–thing spatial relations would change with their changing thing–space relations. Further, in non-Euclidean spaces, doubling in size would differ observably from tripling in size in non-Euclidean spaces – it induces weaker changes in shape than tripling in size. The geometrical facts are familiar enough. But their significance for metaphysics has gone unnoticed. Before I turn to that, I will illustrate the geometrical facts as simply as I can in the next starred paragaphs. The reader who does not need the illustration can skip them.

• It is easy to see what doubling in size comes to in two-dimensional spaces (surfaces) which do not have the character of the Euclidean plane. The consequences generalise fairly intuitively to their counterparts in three dimensions. Let us begin with the surface of a sphere. A triangle in that surface will be any three-sided figure composed of sides which are geodesics of the sphere; that is, great

circles of it. If the triangle is small relative to the radius of curvature of the space, then the sum of its three angles will be close to a
straight angle. But a big triangle will have no such shape. Imagine,
as one side of a large triangle, an arc of a great circle just less than
half the circumference long. At its end points construct two more
great circles to meet at a point polar to this first side of the triangle. The angles at each of the vertices on the equator will be a
right angle (half a straight angle) so together they will sum to a
straight angle. The angle at the pole will be close to a straight
angle, too, so that the whole angle sum will be slightly less than
360° – nearly a complete revolution. That is a gross (but not the
grossest) departure from Euclidicity. This defect (as it is called) is a
function of the ratio between the area of the triangle and the
radius of the curvature of the space. So if a triangle on the sphere
is doubled in size overnight it changes shape. Not only is its shape
different; it now has a shape *it could not have had* yesterday.

- Spaces of positive constant curvature, of which the spherical surface is a two dimensional example, are finite spaces. To double
everything seems to mean doubling the size of configurations of
objects (e.g. stellar triangles) too. But then there might not be
enough space for them to double in. If stars are spread more or
less uniformly all through a finite space, the experiment is not
thinkable. And if the configurations are not doubled, the doubling
just of star sizes will obviously change thing–thing spatial relations.
Some spaces of constant zero curvature are finite too (the torus,
for instance). So even when the local character of space is everywhere Euclidean and doubling is locally indiscernible we can not
guarantee that it will be globally indiscernible. The operation cannot be imagined as performable, no matter what geometry is in
question. We have to know what the space we are dealing with is
like.

- We get defects from doubling lengths in negatively curved spaces,
too, though here the angle sums get smaller as the lengths
increase.

- If triangles change in shape on doubling in these spaces then so
does every other shape. Any polygon can be decomposed into triangles; any closed curve can be approximated by some polygon. As
surfaces change in two dimensions so do the volumes they bound
in three dimensions; and so on up. The result is perfectly general.

- If the space varies in its curvature from point to point, then the consequences of doubling will depend on where the doubling begins and how it spreads. Imagine, first, a surface which varies its curvature from point to point: it will resemble a range of smooth hills and valleys. In some places the curvature will be positive (hill tops and valley bottoms) in other places (saddle backs or hill sides) the curvature will be negative and thus, since we suppose the curvature varies smoothly, some places in between will have zero curvature. Now it will certainly make a difference to thing–thing spatial relations from which centre (whether occupied or empty) the doubling occurs. If one point rather than another is the centre, then some things will be moved further than others by the expansion, and that in turn will determine whether they themselves are doubled in regions of positive, negative or zero curvature. So there will be uncountably many ways of performing the thought experiment and uncountably many distinct thing–thing relations resulting from them. So doubling-in-thought experiments depend on which thing–place relations of spatial occupancy (position, place) are preserved, if any, and how others are altered.

- Perhaps it is worth pointing out that such changes are in general quite grossly observable. How are we to measure the angle sums of large triangles? Here is a crude method. At each angle use a stretched string to draw a circle of standard radius by anchoring one end of the string at the vertex. Now lay the string along the drawn circle with one end of the string at the point where one side of the triangle cuts the circle and mark along the string where the other side cuts the circle. So we measure the length of the arc of a unit circle which coincides with the rotation through that angle. (The length of the chord would serve equally well provided we note when the angle measured by it is greater than a straight angle.) We can transfer the results of this operation from one vertex to another and add the angles simply by adding the arc lengths successively round the unit circle. The one piece of string will give us both the unit circle and the arc lengths. Now if we carry out this operation of adding the two distant angles of the large triangle to the nearby one, we will find that the arcs add to almost a complete revolution (or more for still larger triangles). There is nothing subtle about the defects.

The Leibniz-style Poincaré objection to the realist view of space is fallacious. It is an argument to the detachment thesis – that thing–thing spatial relations are independent of thing–space spatial relations. But the crucial invariance of thing–thing spatial relations requires attachment; that is, precise and unique thing–space spatial relations. Things are to be expanded *through Euclidean space.* Otherwise it will not even appear to legitimate its claims to inconsequentiality. What it presupposes is inconsistent with what it concludes. Note in particular that the dependence of the outcome of doubling on the nature of the space applies with equal force to the special and actually indiscernible case of doubling in Euclidean space. It is not only when doubling will change shapes that we need to specify in which space we are performing the thought experiment. That the doubling operation in Euclidean space has no thing–thing consequences depends on specific symmetries unique to the structure of that space. These symmetries pick out the similarity transformations from the rest. They do not show that there is no structure and no space. They mean that there are specific structures in a specific space.

The string of words 'A is double the size' is not a sentence; it is not syntactically complete. If you like, it is a chimera. To have a thought, true or false we have to know what the thing is twice as large *as.* Just so, the 'thought experiment' given by the directive 'Double things in size overnight' must be syntactically complete. But now it is not enough to know that things are to be double the size they were. We need to know in (through) what they are to be doubled. In each case the chimera is a monster – ill-formed as a sentence or a thought. The imperative which directs us to perform the operation is syntactically incoherent. The 'thought experiment' is unintelligible. We have to specify the space. (Of course we will get general sense out of the idea of doubling things in *some* space, since that is syntactically coherent. But like the idea of painting a thing some colour, it tells us little about how things will look when the job is done.)

Further, we have to double things in space (expand them through it) to hope to support a detachment thesis: else it will no longer be a thought experiment which detaches thing–thing from thing–space spatial relations. The relationist has to begin by giving the realist a space. His problem is how to get this gift back again.

4 Leibniz and the detachment thesis

Leibniz brought several arguments against Clarke to 'confute the fancy' of those who posit the chimera of space. All of them founder on the same rock and by much the same style of navigation as we have seen. The arguments need symmetries peculiar to Euclidean space. The appeal fails without them. (Leibniz ties his arguments to a theological dispute which is beside the point of this paper. Perhaps theological issues warp any straight relevance to modern issues. In that case, what I have to say will concern interpretations of his arguments in the more recent philosophy of science.)

Let us begin with the most important of these, the relativity of uniform motion in Newtonian mechanics (Leibniz's Fourth Paper, §13; Fifth Paper, §52; Alexander 1956, pp. 38, 73–5). How does this engage the inconsequentiality principle? To every body in the universe, whatever its state of rest or motion, add in thought the same velocity vector – the same speed in the same direction. Then everything will change its thing–space spatial relations; everything will be moving through space. Leibniz claims that this will make no difference to thing–thing spatial relations.

Newton's mechanics entails Euclidean space. It presupposes that we *can* add the same velocity vector to every thing. We can do that only if each added vector is parallel to every other. But the existence of parallels is peculiar to Euclidean geometry.

Consider a cloud of small particles which move independently (do not exert significant forces on each other). It is dust. Then, like its component particles, the cloud itself can move at uniform velocity through Euclidean space, free from the action of any force. It will change its thing–space spatial relations (position) continuously but the relative position vectors among the particles in it will not change; they are the vectors which relate particles in the cloud to each other at any time by their distances and directions. So the cloud does not change its shape as it moves. In motion it is just as if the cloud were at rest. That, in turn, is possible because every particle in the cloud can have the same velocity vector. But the idea of the same velocity vector makes sense only if the velocity vector on any one particle is parallel to that on any other. Mechanics allows a symmetry among the relative uniform motions of point particles. That is Galilean relativity. But we need Euclidean geometry to extend this to moving volumes. In other geometries of constant curvature there are no parallels. This may be because all geodesics intersect somewhere (positive curvature). It may be because, given any geodesic G and a point P not on G, there will be infinitely many other geodesics through P, coplanar with G and P, and which do not intersect G anywhere. That is the case when the curvature is negative. Nevertheless, there will be no geodesic in the plane and through the point which is everywhere the same distance from G. It is often said of spaces of constant negative curvature that they have many parallels. This is not true in the sense needed for the thought experiment if it is to yield the detachment thesis. (For a more detailed account of what the motion of dust clouds and elastic solids might be in spaces of constant non-zero curvature see §6.7.)

If a dust cloud moving in non-Euclidean space changes shape, that is not an action of space on things; no more so than the constancy of shape of a moving dust cloud in Euclidean space is an action of space. If the motion of each particle is uniform (as I have been imagining) each will move along a geodesic. That divergence of geodesics in non-Euclidean space is not something which the space does to things which move along them. Space is not a cause and the explanation of the changing shape of the cloud in motion is geometrical, not causal.

Leibniz claims that a uniform motion of simply everything would be without consequences. But this depends on the objects being attached to (being moved through) a quite precise and unique kind of space. Our thought experiment does not show that we can freely detach thing–thing spatial relations from thing–space spatial relations. The relations depend on each other. The argument to the thesis is fallacious; it presupposes (tacitly entails) the attachment it attempts to deny. Here again, quite clearly, the attachment is presupposed not just in the case of non-Euclidean spaces where uniform motion is well defined and observable. The inconsequentiality Leibniz needs is a symmetry of Euclidean space. He needs the symmetry to be concretely there in something real and not chimerical.

This suggests that Leibniz simply assumed that Euclidean symmetries were inescapable. But it is not quite so simple. He seems to have thought he could derive the symmetries just from the indiscernibilty of spatial points one from another. His 'proof' runs as follows.

Space is something absolutely uniform; and, without the things placed in it, one point of space does not absolutely differ in any respect whatsoever from another point of space. Now from hence it follows (supposing space to be something in itself, besides the order of bodies among themselves) that 'tis impossible there should be a reason why God, preserving the same situation of bodies among themselves, should have placed them in space after one particular manner, and not otherwise.

(Fifth paper, §5)

The passage does not make it clear whether Leibniz thinks the uniformity of space derives from the indiscernibility of points or vice versa. But we should not accuse him of taking the symmetries for granted simply from Euclid's geometry.

In any event, his conclusion does not follow. Spatial structures richer than topology are not gained by adding to properties of points. Affine and metric structures are properties of the space as a whole and 'emergent' relative to properties intrinsic to points. The presence, absence or variation of curvature is not an intrinsic difference among points. It depends on the structure of the space. Curvature is a limiting property of points, deriving from the (possibly varying) structure of the space round them.

Leibniz also tells us that it could make no difference to thing–thing spatial relations if everything were displaced some arbitrary distance and all in the same direction. As before, the phrase 'all in the same direction' requires parallels. But there are new problems. He tells us that there would be no change to thing–thing spatial relations if things were interchanged east and west (Leibniz's Third Paper, §5; Alexander 1956, p. 26. Whether this is supposed to be the result of a reflection, or a rotation round an axis makes no odds). These displacements are not to be thought of as processes of moving things to new places. Leibniz's theological metaphors make the point clear. It would make no difference had God created things not where he did, but so that each came into being 100 metres further north or with east and west interchanged. (Put aside proper worries about north, east and west as the same directions everywhere.) The point is that the displacement would yield differences if the curvature of the space varies from point to point. Indeed, if our thought experiment asks us to imagine perfectly rigid bodies, not even God will be able to perform it. A rigid body cannot be relocated in a region where the curvature is different, since no shape in a region of one curvature is the same as any shape in a region of different curvature. An elastic body might be relocated by distorting it, but a *perfectly rigid* body cannot be distorted. God cannot act a contradiction. Among elastic bodies, we can expect rubber objects to distort more than steel bodies, and expect that it would be easier to relocate them. But we can detect and measure these various distortions. We are not dealing here with Reichenbach's universal forces, but with common or garden elasticity, tensile strength and the like.

The inconsequentialities of Euclidean space are not the product of a confused hypothesis about a non-entity. They are the symmetries of a unique and well-defined entity with a perfectly definite, if bland, structure. Change the structure of that entity and you have the thing–thing relation consequences which Leibniz thinks impossible. We cannot argue as he does with the inconsequentiality thesis as premise. It is false.

5 Some conclusions so far

The detachment thesis lacks support. Any version of the reductive relationist program for banishing space from our ontology which depends on Leibnizian arguments for the detachment thesis is circular. The relationist thinks that his hardest job lies in completing the reduction. But that is really the easy part. The relationist singles out a subset of spatial relations, those which hold among the occupied points of space (or spacetime). The realist regards this as entailing the singling out parts of space (such as the paths across which two things are at a distance, or volumes of space in which the matter of things is disposed) which cannot be detached from the holding of the spatial relations. Put in that light, the problem of completeness is simply the problem how to determine the character of a whole space by specifying relations among certain parts of it. It is by no means a contemptibly easy problem (see Mundy (1983). But it is not obviously a metaphysical one either. Nor is it the problem which realism is concerned to say is insoluble. The vexed question lies elsewhere in the use for reductive purposes of spatial relations detached from the parts of space (paths etc.) which the realist takes as needed to sustain them. In short, the metaphysical issue is not to deduce the structure of the whole from only some of the parts. Can we detach thing–thing spatial relations from thing–space ones quite straightforwardly? The prominence of Leibniz-style arguments in the literature suggests that, *on the face of it*, we can't; these arguments aim to show something that needs to be shown. Suppose that it can't be shown. Then, as ontologists, we should be no less worried by the nature of spatial relations among occupied points than we are about any other spatial relations. These relations carry with them the mediating parts of space. The nature of these parts (volumes, paths, points) is metaphysically no different from the nature of the whole. Whole and parts equally have the same awkward features listed in §6.1. The real problem is to show that thing–thing spatial relations do not depend on thing–space relations. Relationism without a detachment thesis to extricate spatial relations from dependence on parts of space lacks reductive teeth.

Leibniz appeals to the Indiscernibility Principle; it is dubious. However its converse is highly plausible. We might call it the Discernibility Principle and express it as follows:

DP: Consequential properties are real properties.

This is the contrapositive of the converse of IP. We can gloss 'consequential property' after the style of 'inconsequential property'. Suppose we adopt DP. Then we have really a rather plausible argument that position in space and the quantity of space filled or occupied by a material thing are real relations. So some thing–space relations are genuine and not chimerical. In general, change in these relations changes thing–thing spatial relations in consequence. If the latter changes are real, so are the changing properties that induce the consequences. But then the thing that has the properties must be real too.

Thirdly, and more generally (but more vaguely), Leibniz's conclusion has been our paradigm for a whole range of epistemological claims about the probity and point of concepts in science and outside it. The path for many of these has been the easier because of the apparently brilliant success of their Leibnizian forerunners. Reductive epistemology did not begin with Leibniz, but his incisive attack on Newton wrote a distinctive new page. This, together with the fact that classical physics after Newton seemed to endorse his attack, lent a respectability to reductive empiricism, in one form or other, which it would hardly have gained otherwise.

Theses that something or other is unobservable in principle have proved vulnerable. Most of them were proposed as having a status unhappily like that of synthetic necessary truths, though, like the verification principle, used to deny that there could be any such truths. They are as fragile as allegedly synthetic necessary truths deserve to be. Yet they have motivated various draconian principles by means of which seemingly respectable conceptual structures were run out of town. This vigilante epistemology has seemed judicious largely because Leibniz's dazzling arguments about space supported it. If Leibniz's arguments are our paradigm for the probity of inconsequentiality and reductivism, their failure might become our paradigm for impeaching them.

Thus the role of epistemology in metaphysics has been regularly misunderstood and overstated by confining ontology to a dependent and sharply limited role in a puritanically conceived theory of knowledge.

In this mistaken and destructive idea of things, a fixation on Euclidean geometry played no small part. It is ironical to reflect that it is the very simplicity and symmetry of Euclidean geometry which has misled us. The simplest possible spatial world is confusing in its very simplicity and blandness.

6 Further assumptions and consequences.

Ontologically, Euclidean space is surely just the same kind of thing as non-Euclidean space. Newton's theory demands that we include Euclidean space in our ontology. The metaphysical nature of this commitment is no more disturbing than when our theory commits us to a more readily observable non-Euclidean space. No doubt Euclidean space is bland. But that cannot change the nature of an ontic commitment to space. Non-Euclidean space has all the disconcerting features in our list of §6.1 just as surely as Euclidean space has them. So I make the following assumption:

A1: Structural symmetries do not affect the metaphysical nature of an entity.

However, the structural symmetries of the entity may well alter the way it enters into our theories. We now see that it is the realist who can postulate spatial symmetries and thus explain why, in Euclidean space, we can never observe the difference between doubling and constant size, between rest and uniform motion and so on. The realist owns the symmetries, not the relationist. It is the realist's realism that tells him (and tells us) why we cannot care about the difference between one of these states and another. In fact, it is the relationist who has no clear, coherent, non-circular account of the relativity of motion. That relativity rests, not on nothing, but on symmetries of a real entity, Euclidean space. *Therefore the realist does not have to answer the objection which is standardly brought against him.* The objection is that the realist is obliged to make these inconsequential distinctions and must take them seriously whereas the relationist need not. But, on the contrary, the realist is just the person who need not concern himself with them since he and he alone has a clear picture of their source and their nature and their lack of consequences. He alone has a coherent account of when they hold and how they may fail. They spring from the character of Euclidean space. How can anyone oblige him to care about what *the nature of his postulate* makes plain to him that he *cannot* care about?

Let us use 'absolutism' to describe the view that there are some symmetry-transcending properties that matter for science or for metaphysics. Then realism does not entail absolutism. Rejecting absolutism takes no step towards relationism. Newton was an absolutist.

None the less, motion is motion-through-space. '*x* moves' is good grammar for we can say whether it is true or false if it is contextually clear what the tacit relatum is. But the deeper syntax of sentences of motion requires that the referent be filled out on demand – the demand of how the sentence is to be understood. Now the non-Euclidean cases make it plain, first, that the demand *need* not be met by giving an observable thing as the tacit relatum. We may say instead whether or not things move in a non-Euclidean space. But that is not all. Familiar motion, relative to other things in space, does not yield an equally significant idea. It is a weak or recessive sense, bound to yield to the deeper more general requirement that a thing is in motion through space. Thing–thing relative motion might turn out to be a symmetric relation in a suitable theory. But the deeper thing–space motion is asymmetrical on semantic grounds. Space does not move relative to things which it contains (but see §6.7, objection 1).

The realist can draw a clear map of the path followed by the post-

Newtonian classical physicists. He can relate rest and motion to any of the frames of reference which dynamics and the symmetries of Euclidean geometry together select as indistinguishable. It is at least as clear a map as a relationist can draw. It does nothing to undercut his commitment to a space with indiscernible symmetries. The relativity of motion rests on the commitment. Every frame of reference defines a Euclidean space with the metaphysical properties noted in our list in §6.1. We may, if we wish, consider these as different hypotheses about what is at rest, what in uniform motion and the like. Our realism tells us that the difference between these hypotheses cannot matter. There is no clear position which allows us to say that space forms no part of the ontology of post-Newtonian physics. That ontic commitment is there in every choice of frame. It is unclear how relationism avoids begging this question.

I conclude that realism does not entail absolutism.

However, there is a wild card in this deck. It is the metaphysics of continuants. Continuants have spatial parts but not temporal ones. They have histories, but the history of a continuant is a property of it, not a part. Suppose that you and I are continuants. Then I am the person who wrote the paper you are reading: there is no paper-writing part of me nor paper-reading part of you. Neither of us is extended in time. Our histories are our tensed properties, not parts of us. Newton supposed that space was a continuant (though not a substance) and, for this reason, was obliged in general to admit that questions of the identity of places at different times were meaningful and admit that there were different indistinguishable hypotheses to choose (freely) among. A realist need not follow this path. Instead, place classical mechanics in spacetime; this will mean that questions about which is the same place at different times cannot arise because spacetime is not a continuant entity. Nothing in this bows to detachment theses or spatial relationism: spacetime causes much the same consternation among ontologists as space causes, and for much the same reasons.

But neither that nor the fact that a realist sees he sometimes cannot care about distinctions between rest and motion means that the questions about rest and uniform motion through space are meaningless. They are idle in the case where space is Euclidean, but only just. Give the space a little uniform curvature and the distinctions engage at once with thing–thing relations. Surely the limiting case of zero curvature inherits the sense of the range of cases of which it is the limit? It is not clear why not. Suppose a space that is changing its curvature from (everywhere constant) positive to (everywhere constant) negative curvature. Do the questions not retain both their point and their meaning as the changing space passes through the limit between the two? I assume that it does.

Suppose we deny the assumption A1. Instead we assert:

A2: We cannot be ontically committed to a space with symmetries.

It would seem to be IP, the indiscernibility or inconsequentiality principle which might motivate A2. But if that is right, then it would seem to commit us to the view that though there might well be non-Euclidean space (best, perhaps, space of variable curvature) there could never be a Euclidean space. It is too symmetrical, observationally bland or anaemic to exist. J. L. Austin once spoke of the idea that existing is like breathing, only quieter. We seem to encounter something like this here, but without the quietism.

Perhaps it is this view which John Earman recommends to a reconstructed ahistorical Leibniz in almost the only place in the philosophical literature where the ideas I have been raising seem to have been mentioned (1970, p. 303. See also, notably, Stein 1977). The relevant two or three sentences come in what appears as a casual afterthought in his splendid paper. Earman points out that Leibniz's arguments fail if the space lacks the relevant Euclidean symmetries. He grants Leibniz anachronistic knowledge of the variety of geometries, and imagines him objecting that Newton used too bland a space to allow him to claim its real existence. Or so I understand Earman. The brief passage is acknowledged and repeated, almost verbatim, in Michael Friedman's excellent (1983), p. 221. This view of things seems to portray Leibniz as insisting on A2 and going on to claim that Euclidean space could not exist, though a non-Euclidean space might.

But I find this suggestion unacceptable. It confuses geometrical differences with ontological differences, or blandness with non-existence. If Leibniz had indeed raised this objection. surely a decisive reply would have been available to Newton. For, first, neither of them had the least *empirical* grounds for thinking that a non-Euclidean geometry described the structure of space; second, Newton would rightly have claimed that Leibniz was attempting to saddle him with a needlessly complex geometry in order to satisfy an obscure maxim of epistemology which has no place in natural philosophy. Surely the conclusion that there could be non-Euclidean spaces but there could not be a Euclidean one is absurd (though not contradictory).

A2 is the child of IP, the inconsequentiality thesis; it leads to absurd consequences so any version of IP which entails it must be false.

Further, Leibniz's argument is fallacious. It is not just a matter of the falsity of his tacit premise – that every possible space has symmetries of Euclidean space. The light shed by non-Euclidean geometries merely throws into sharp relief an already present appeal to a symmetrical space. Leibniz appeals to that symmetry throughout and it would none the less belong to an entity of the sort he wants to deny, but has to presuppose, even if every possible space possessed it.

Finally, I assume that realism is ill-described as substantivalism. Newton said that he was not a substantivalist and gave a clear reason for saying so. Space has no action on things. Another idea of a substance is simply that of a continuant. Neither of these is entailed by realism, but only by specific forms it might take. In General Relativity, mass and energy are features of the spacetime metric and affinity. I find it strange that this is described as a relationist view of space and opposed to what is called substantivalism. What have relations to do with it? What else would one want to describe as substantivalist other than the opinion that spacetime stores energy and mass, that particles may be generated out of and decay into spacetime structure? Certainly not the view that *is* described as substantivalism – that the manifold, despite being undetermined in its structure by the matter tensor, is real. The manifold is too attenuated to be cast as a substance. But it is nevertheless arguably real and the view that it is is surely best so described – it is realism, not substantivalism. Of course any substantivalist view will make the distinction between container and contained less straightforward than Newton supposed. But that is no feather in the cap of traditional relationism (see Earman and Norton, 1987).

7 Replies to some objections

1. Let us reconsider nocturnal expansion. When we performed the thought experiment, we did not double strictly everything. We did not double the space. But if we double the space as well as the

objects, then there will be no change in thing–thing spatial relations. So is not this Leibnizian argument valid after all?

No. The argument has to yield a particular conclusion, the detachment thesis. It is that thing–thing spatial relations are independent of thing–space spatial relations. But the revised thought experiment does not detach thing–space spatial relations from thing–thing ones. It preserves them. Since it is not a thought experiment in which there is a detachment, it is powerless to defend a detachment thesis.

Perhaps this is a superficial reply, because we have given no meaning to the operation of doubling the size of space. We have yet to give it even a decent syntax – make a well formed description of it. We have not said through what the space is to be doubled. Before we can give even the meagre 'sense' of well-formedness to doubling our space as well as the things it contains, we must suppose that there is some higher-dimensional space that contains it. But a higher-dimensional containing space for our space is just an idle fancy. It plays no role in Newtonian mechanics or in any other actual theory of ours. To give to doubling the size of space something we could fairly call a semantic meaning (something stronger than well-formedness) the higher-dimensional space has to be given some point in the theory within which discussion is going on. We have yet to see that point. So the idea of doubling space as well as things lacks meaning – so far.

But even if we made the fancy less idle by giving some point to the higher-dimensional space, it does not follow that doubling our space through this higher-dimensional space would be inconsequential. That depends on whether the wider space is Euclidean. One obvious way to give speculative point to the wider space is to make it non-Euclidean, for then the doubling of our space would change both its shape and that of the objects it contains. The doubling would be consequential. We are still fixated on Euclidean space if we suppose that doubling the space will change no thing–thing spatial relations.

It might seem that the objection has no need of this further space, as my colleague Michael Simpson pointed out to me. The curvature of the space is intrinsic to it, not dependent on some higher-dimensional containing space. So it seems that we might double the radius of curvature (i.e. halve the curvature) without referring beyond the space to another which contains it. Then the objection makes sense. However, we can all agree that metrics equivalent up to a scale factor are not meaningfully different even when the space is finite because of its curvature. This may look like an agreement among relationists and not something to which I can happily turn here. But the point is syntax (addicity) as before: a scale factor can be given sense only when we have something to which to link our space by such a relation as 'six times as large as . . .'. We can make sense of something being as large as it was only by reference to the quantity of space it filled. As I see it, a relationist ought to insist on this just as firmly as a realist does. The further debate is whether anything could fill the bill as a relatum unless it is somehow privileged (observable, changes thing–thing relations etc.). We all agree on a point about well-formedness: doubling the space makes no sense as an intrinsic non-relative operation. The dispute is about whether and how we might give relative sense to it. If the hypothesis of a containing non-Euclidean space through which our space is doubled has consequences for thing–thing relations, the relationist will have difficulty in disallowing that it is meaningful. I am suggesting that whoever raises the spectre of somehow doubling the space is, once again, giving me what he hopes is the empty hypothesis of a containing space and defying me to make use of it. But the use is clear.

It is much the same if our Leibnizian fancies fly to the bizarre height of adding to the space of our objects the same velocity vector we added to the objects themselves. Now when everything moves at a uniform velocity, our space moves too, along with the objects. The fancy is well-formed only if space-and-objects, holus-bolus, moves through a wider containing space. But, again, if this wider space is non-Euclidean then both our space and the objects in it will change their shapes, for they will have no parallels in the wider space for the points to move along. The appeal to space as among the things to be changed by any of these thought experiments gets us no closer to the inconsequentiality Leibniz needs.

Contrast this idle fancy with something else. General Relativity is a theory in which tensor equations describe the physical fields. The fields are determined by quantities that might roughly be described as observables. They fix the metric, the affinity and other things. But they do not determine the lie of these tensor fields on the basic manifold. They do not even fix how the fields lie in local regions beyond which everything is specified. That is the essential thrust of what's become known as the hole argument (Earman and Norton 1987. See the discussion of this in §9 and its description in §9.1). Nevertheless, the manifold is not like the idle fancies just discussed. Although the symmetries of the manifold may give us a *motive* for wishing to rid ourselves of it in our ontology, it provides no sort of argument which shows us how to delete it. We cannot define tensor fields except on manifolds: books on tensor analysis begin with manifold theory, as a casual look at chapter headings will show. The same is true of many books on General Relativity. So we cannot begin a tensor theory without first postulating a manifold. General Relativity leaves us with a manifold that has symmetries permitting a variety of different placements of the tensor fields within it. The symmetries show us that it cannot matter which lie on the manifold we choose. The results in terms of observables will not change. But this does not mean that we can take our postulate back and give up realism about manifolds. For, here again, it is in the structure of the manifold – from which the theory must begin – that the symmetries lie. They cannot lie in a nonentity. Non-existence has no character – in particular, not a symmetrical character. The hole argument fails as an objection to realism. But, again, the realist's perception of the symmetries in the entity he postulates leaves him perfectly free to care nothing about these differences. He plainly sees from the nature of what he postulates that he cannot care about them.

This point is worth making because it may help us to formulate more useful principles about what is meaningful and, rather differently, what to care about in scientific or any other sort of theories. This is a really difficult task. The IP is visibly too strong. Yet we do not want to admit as meaningful every sentence containing some terms semantically anchored in their theoretical role. The preceding example of doubling physical space relative to an idle space meets that weak condition. 'Space' is a semantically anchored term. The idle space is a space in the same sense as physical space, so the sentence about doubling space is syntactically well formed, and thus it is weakly in order semantically. Until we give the idle space a role in theory we ought to deem the doubling-of-space sentence meaningless, though open to acquiring a meaning. For who can say how theory will advance next? The case of the manifold is different: it certainly has a role in General Relativity. Let those who deny it show us how to get along literally without it, instead of merely pointing out its symmetries. But I shall content myself here just with the easier task of concluding that there is much work still to be done in this regard.

2. Here is another objection. Nothing said so far takes us beyond the observation of thing–thing spatial relations. These are all we have under the hypothesis of non-Euclidean space to show us that things would change if doubled over night. Even then, the observations are not enough to establish that the space is non-Euclidean. Poincaré, Reichenbach, Grünbaum and others have shown with ample clarity that no observations of the sort envisaged can suffice to commit us to a space with any special character whatever.

Reply: our central question is the argument for the detachment thesis. The issue concerns this conditional: if we change thing–space relations there are no consequential changes for thing–thing relations. This conditional thesis has been refuted. The objection challenges us to defeat the converse conditional: if thing–thing relations change as envisaged, then new thing–space relations must have produced the changes. But we have no need to refute this. It is not the detachment thesis.

Yet we could refute it. At least, we could argue against it in perfectly familiar ways. The objection throws gratuitous obstacles in the path of reasoning from the evidence to the best explanation. That sort of problem is in no way peculiar to arriving at good geometrical explanation. That is a perfectly general problem for scientific theorising. We are too familiar (from Quine and Duhem) with the way in which tinkering with auxilliary hypotheses and initial conditions can make any body of evidence consistent with ugly, arbitrary, deviant theoretical conclusions. Though the metaphysics of space has proved a happy playground for the invention of obstacle courses, there is no reason to suppose it suffers in a more intimate way from the general malaise. The DP together with the nature of non-Euclidean geometry suggests that the hypothesis of space can do genuine explanatory work (even though it makes no appeal to the action of space). What we have to do here is to argue for the best explanation of the phenomena. Though such arguments are not formally valid, they convince rational people.

Given the evidence we have been speculating about, inference to a space with one or other definite geometry looks like the most elegant and fruitful explanatory hypothesis. We know how to treat space (or spacetime) as something which has concrete relations to concrete things, and how these changing relations might explain observable changes in the world. General Relativity does this for spacetime. But geometrical explanation of concrete changes by means of concrete spatial relations is clear too (see §6.7 for illustration and discussion of this). As far as I can see, the main reason why we hesitate to accept geometrical explanations in these cases is that we think the idea of something immaterial and imperceptible – space – is absurd. This does not look like a strong reason. The idea of the material has undergone drastic changes in physics over the last century; the idea of what is observable is anthropocentric and local. It has usually been held that the postulate of a space with a well-defined geometry can do no explanatory work. But it plainly can. Our sense of uneasiness about it begins to look like a vulgar human prejudice.

Is it coherent, meaningful, or intelligible to commit oneself ontologically to an entity which has the sort of stressful combination of properties listed in the introduction? Surely the answer to that modest question has to be 'Yes'. It is hard to see what sort of objection can be seriously raised against the intelligibility of entities like space unless it is that space is neither perceptible nor material. Of course, spatial kinds of things are not perceptible and generally not material. But it is not clear why these should be regarded as defeating the commitment as long as space has well-defined characteristics, is a particular, has concrete relations to concrete things and can explain concrete changes in things (such as change of shape after doubling). If space can deliver these and other sorts of theoretical goods for us, it does look like prejudice to insist on materiality and perceptibility. And clearly Leibniz shows us how well we can understand the characteristics of a space by appealing so ingeniously to a Euclidean character for it.

Could it ever be useful to include space in the ontology of some theory or would we always be better off trying to avoid the postulate? The answer to this question is much less clear, though my own view is that space does some work in easily imagined theories, and does it elegantly enough to earn its keep. Relationist alternatives, by comparison, rob theoretical language of much of its assertive force and burden it with awkward conventions and indeterminacies. They impede a clear view of what the theory does, clutter it with non-factual elements while offering no gain beyond the economy that they do not postulate space. Even then, I have been arguing, it is unclear how they really do avoid the postulate.

Leibniz's argument has loomed large because relationism needs to argue for the detachment thesis. The challenge to Clarke was no mere historical episode in which we now take only an antiquarian interest. It remains the most attractive relationist gambit. Relationism has no point without the detachment thesis. The argument of the challenge is fallacious, begging the question. On the other hand, if the Discernibility Principle is true, we have a strong argument from it to the meaningfulness of thing–space relations generally and so of space itself. It is easy to see how we might arrive at a posit of non-Euclidean space as our best explanation of what we observe.

7 What can geometry explain?

1 Introduction

If space and spacetime are explanatory ideas in physics and common sense, just what sort of explanation do they offer? This question can't be answered until we have a clear idea of what kinds of states, events and processes we might be able to explain by making use of space and its structures – of geometry, in short. It is not obvious that space does have any significant role in explanation, but perhaps this can be made more obvious by exercising our imaginations about how things would be and what would explain them if space were very different from the way we find it. Part of this essay considers that question.

There is a central question about whether explanations which draw on the nature of space are causal or of their own kind. Conversely there is a question whether we can discover what the nature of space or spacetime is unless we find it out by means of the causal influences revealed in what it explains about the fixed properties of things and the occurrences of events. If we look at this question outside the context of GR, which is what I do for most of the essay, there is good reason to reject the tie between what geometry can explain and what we explain by means of causes. There are advantages in a clear sight of this. The situation in GR is more ambiguous and I touch on it only rather briefly.

I turn first and briefly to an example of the explanation of a fixed property of familiar things.

2 Hands and handedness

Consider the familiar property of left–right differences. Since we take
our hands as displaying this difference, it is a geometric property of
very familiar things indeed. Since I have discussed this at length
(Nerlich 1994, §2), I will be brief here.

The handedness of an *n*-dimensionally asymmetrical object in an *n*-
dimensional space depends on a global topological feature of the
space called orientability. This is best grasped by seeing it as an aspect
of the shape of space. Two-dimensional examples illustrate the general idea. An L-shape is two-dimensionally asymmetric if we embed it
in the two spaces of the Euclidean plane, the cylinder and so forth.
But it is not handed if we embed it in the space of the Möbius strip or
of Klein's bottle. In these spaces an L-shape carried rigidly round the
space comes back locally congruent with its former reflection. It is
homomorphic, not enantiomorphic or handed.

What is being explained here, and how does the explanation work?
Ordinarily, I suppose, we take the difference between left and right to
be a primitive, simple one which we can gather directly by inspection
and which is to be defined ostensively (see, e.g. Earman 1971 and
Bennett 1970). The first part of the explanation points out that this is
a mistake. Asymmetrical objects may be enantiomorphic or they may
be homomorphic depending on the rigid motions (paths) which the
space provides for them. Up to this point, we are explaining the concept of enantiomorphy by analysing it in terms of asymmetry and
orientability, which is a property of space globally. Whether or not
space has this global property is not a question of any change which
the space enforces upon objects in it, nor does it sustain the handedness of objects by some action it performs upon them or some disposition it has to make things handed (or not). It is a geometrical question about which pathwise connections are in the space. Space *closes
off* no paths to objects. If it is like anything outside geometry, orientability is like an existential condition on space; simply *there are not*
the paths in such a space which permit asymmetrical objects to be
homomorphic. But whatever one may make of that way of classifying
what the explanation is telling us, I take it to be plain enough that it is
telling us nothing causal.

Another question might suggest itself at this point: cannot this
explanation be reduced to something relationist after all? This was

the theme of a rejoinder by Laurence Sklar to the first version of my paper about hands. But Sklar's method of dealing with the problem of enantiomorphy is itself both absolutist and circular, or so I argued in reply. (1994, §2.8). It is circular because when Sklar talks of possible motions, the sense of 'possible' intended has to mean 'permitted by the pathwise structure of orientable (or non-orientable) space'. It is absolutist because motions are in fact *paths*, i.e. parts of *space*, and of the same ontological order as it. But I will not pursue this argument again here.

I conclude this section by claiming that we might well discover whether space is orientable or not (if space if finite) even though its orientability (or lack of it) does not *cause* things to be enantiomorphic or homomorphic. Orientability being a *global* property of space, it evades attempts to construe it as a causal one, since it is not pointwise definable. It does not vary from point to point nor does it stay the same.

3 Constant curvature and what it explains

What happens when we apply classical physics to motion in spaces which have non-Euclidean geometry?[1] Let me simplify things by assuming that the laws apply to strict point particles. My first example assumes only such an application of the first law. Any affine or metric space has a pointwise definable property Ψ of curvature. I will be arguing in this section that, when Ψ is non-zero and constant throughout the space, this makes a difference to the behaviour of matter. Though classical mechanics predicts the difference in the state of things, it is plainly not the case that the constant curvature of the space *causes* the characteristic states of matter. I take it that, in the context of mechanics, we can speak of causes only when we can speak of forces. I hope to make it quite clear, intuitively, that there may well be a causal story to be told in these circumstances about the behaviour of matter, but that it nowhere causally involves space or its curvature. However the curvature of space is certainly needed for a wider style of geometrical explanation as to why things happen as they do.

---

1 I am indebted in this and the next section to Professor Angus Hurst, Mathematical Physics Department, University of Adelaide.

The two dimensional spherical surface is a space of constant positive curvature. Further, it is intuitively rather obvious to us what its geometry is, so I will begin by looking for some analogies there. The straightest paths that lie in the surface of a sphere are its great circles. If we apply the first law of motion to material points moving on a spherical surface, then it tells us that the force-free motion of a particle will be at uniform speed along a geodesic, which is one of the great circles. The figure shows us the spherical surface and a geodesic *E*. At various points on the geodesic are vectors (arrows) in the surface of the sphere each of which is orthogonal to *E*. Clearly, if these vectors are moved along their own lengths they will follow geodesics all of which meet at the point P which is polar to *E*. Thus, there are no parallel geodesics in the two dimensional surface of the sphere.

Now consider a cloud of unconnected particles moving at the same speed in a common direction in Euclidean space. Each material point continues to move parallel to every other at the same speed and the cloud will maintain its shape indefinitely. The velocity of the cloud can therefore be represented by a single vector through its centre. Its momentum, equally simply, is given by the same vector, up to a factor of mass. But it is seldom so simple in non-Euclidean spaces, and clearly more complex in the standard simplest cases of constant curvature. The two-dimensional case of constant positive curvature, the sphere, makes this obvious, and similar results extend to negative cur-

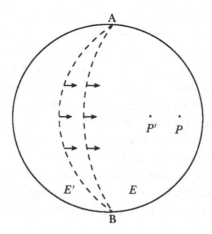

Fig. 7.1. Vectors orthogonal to geodesics which meet in the same pair of polar points A and B.

vature and to three dimensions. There are no parallels on the sphere. So, as is again clear from fig.7.1, a patch of unconnected particles moving at the same speed through the 2-space of the spherical surface can move in the same direction only in a weak sense. That is, at some particular time, t, the vectors of all the particles are orthogonal to members of some family of geodesics which meet only in the same pair of polar points. E and E', in the figure, are members of such a family; particles in the patch whose velocity vectors at t are orthogonal to E or E' converge on P or P' respectively at some later time t'.

The first law of motion entails that these particles will follow the geodesics in which their velocity vectors lie. As we saw, these all come together. Hence the patch of particles will change its shape with time. Clearly, we can no longer deal adequately with the mechanics of the cloud of particles by means of a single vector through its centre. We must treat it as a vector field. So momentum, too, is a vector field, identical with the velocity field up to factors of mass. These fields will be much less simply summed than is possible in Euclidean space. Generalising again to higher dimensions and negative curvature, it follows that a cloud of dust inertially moving in non-Euclidean 3-space of constant non-zero curvature will change shape and volume over time.

Here, then, is one example of an observably changing state of matter which involves no causes at all, since the motion involved is purely inertial, governed just by the first law. No forces operate at all. The curvature of space explains the change of shape in the context of classical physics but, quite clearly, is nowhere causally involved in it.

Analogous results hold for the motion of dust clouds in the majority of 3-spaces with variable curvature, and I draw from them the same conclusions about the geometrical, non-causal, style of explanation which spatial curvature gives.

The general problem for a swarm of interdependent particles moving in spaces of non-zero constant curvature can be understood just by looking at the motion, in two dimensions, of an elastic membrane moving through a spherical surface. Classical physics again entails that each point in the membrane will move along a geodesic unless acted upon by a force. But now the elastic forces in the membrane will act on its individual molecules so as to resist change in the membrane's shape. Elastic forces are electromagnetic and hold among the molecules of the membrane, so that stress in the membrane may be regarded as a vector field. Clearly this vector field interacts with the vector field of momentum, so that the outer molecules have their momentum changed just by stress in the membrane which will be in a constant state of tension as it moves. The stress energy is acquired as the object is accelerated, but since the curvature of the space is constant the total field of an inertially moving object need not change with time, so that the body as a whole will satisfy the principle that momentum is conserved. The mechanics of the motion is rather analogous to the familiar case of an elastic solid rotating in Euclidean space which conserves angular momentum under stress. We never say that space causes this stress. But it still defines the force-free trajectories and thereby explains how the internal elastic forces work to pull the matter round and away from these paths. The structure of space plays a clear explanatory role here.

Here again, the results extend in quite analogous ways, to spaces of three dimensions and constant negative curvature. In these dimensions, stress is a tensor field and this adds some complexities to calculations which need not detain us. Quite generally, the mechanics of the inertial motion of elastic areas or volumes in spaces which have no parallels must be quite different from the mechanics of their inertial motion in Euclidean space. But even in the simplest cases of constant non-zero curvature, an inertial uniformly moving body is under stress. Moreover, its stress is clearly a function of its speed and of the curvature tensor Ψ. Thus its motion will be absolutely detectable from its internal mechanical state, just as for classical physics rotation is detectable in Euclidean space. So the inertial motion of elastic volumes is not relative in such spaces; hence motion in general cannot be.

It is clear, I think, that it would be quite wrong to claim that space plays a causal role here in stressing the body. The causal story is exhausted in our account of how the vector or tensor field of stress interacts with the momentum field. It tells of electromagnetic forces acting among molecules so that some are accelerated and their momentum field is changed. That assigns a clear role to the first and second laws of motion. The third law is clearly fulfilled in action–reaction pairs among molecules. Space no more enters the picture causally than it does when classical physics explains the mechanics of a rotating elastic solid. It doesn't push; it doesn't pull; it exchanges no energy with the particles or bodies. Galileo's most liberating breakthrough in his revolutionary treatment of mechanics was just to point out how much sense it makes to regard uniform motion not as a causal phenomenon, but as 'natural'. This is equivalent to regarding geometrical explanation as non-causal, too. To speak of cause here is to trivialise the false but interesting thesis that all explanation of events is causal. It evacuates it while it rescues it. However, that space has constant non-zero curvature plays a crucial geometric role in explaining why and how much the thing is stressed in its motion. Thus it would also follow that the constant zero curvature of Euclidean space plays a role, neither trivial nor causal, in explaining how bodies move in it without stress.

4 Variable curvature and what it explains

My next example of geometric explanation makes reference to a variable pointwise definable property of space. I want to look at explanations which use the curvature of space, and I assume that this feature will differ from point to point. I will look at how classical and SR physics can be applied to the motion of an elastic solid through a space of variable curvature. Again, I hope to show that though causes play a part in explaining the oddities of motion in these circumstances, the complete explanation contains a part which is clearly not causal but recognisably geometrical. I will use the concepts of momentum, energy and elastic (electromagnetic) force to show as before that the relevant causal interactions are all among material particles. Space plays no role in them other than to define the directions and distances which they act across.

There's a proverb which says that square pegs won't fit in round holes. If the peg were made of soft rubber it might be made to fit the hole if we were to exert a force on it, change its shape and induce some stresses in the elastic material which it is made of. We will not need a distinctively geometrical style of explanation to show us what is going on here, since we will not need to make use of any properties of space itself. But suppose our peg is a Euclidean cube of soft rubber and our hole an empty region of non-Euclidean space. We will just as surely have to change the shape of the peg if we are to move it into the hole. This means that we must exert a force on the cube and stress it, as before. But here we cannot exert a force against the constraining walls of the hole, since it has no walls. The force is spent wholly in changing the peg; but this is something we need to look at in some detail. Fortunately, we can keep the details qualitative and largely intuitive.

Let us begin by looking for some analogies in two dimensions. Suppose we have a square piece cut from a flat elastic membrane, that is, one which will lie flush on a Euclidean surface without internal stresses. Stresses are elastic forces between parts of the material which a thing is made of and are basically electromagnetic forces. This square piece cannot lie flush with a spherical surface unless the area of the sphere is very much greater than the membrane's area. On the sphere there are no four distinct geodesics which bound an area and which intersect orthogonally at all four corners. If the piece is to fit flush it must somehow be stretched out in some parts or squeezed up in others. We must change the ratio between its perimeter and its area since no part of the area of the sphere has such a perimeter/area ratio as the piece of membrane takes up in its stress-free state.

Clearly, in the light of these facts, if we move a square piece of membrane across a surface of variable curvature on which we make the membrane lie flush, then it cannot move freely. Let us suppose that frictional forces can be neglected. In that case, what impedes the free sliding of the membrane across the surface can only be that its perimeter–area ratio must be readjusted to match the differing curvatures in regions of the surface itself. We must use force on the membrane to stretch or shrink it before it can fit flush against new parts of the surface. This provides a simple analogy of what we will find in three-dimensional cases of variable curvature.

As before, the examples are explained in detail by considering how the momentum field changes and is changed by the tensor field of stress. Spaces of variable curvature are too diverse to allow a general intuitive argument directly about vector and tensor fields which would give a useful result. But, as the case of the flat membrane on the sphere shows, once the geometry of a space changes from region to region, so the disposition of material within an elastic solid must change as it moves through the space. Thus a volume of soft rubber which is in a stress-free state in a Euclidean region cannot be in that shape if moved into a region of non-zero curvature. Just as there is no such shape as a square in the space of the spherical surface, so there is no such shape as a cube in these non-Euclidean spaces. Therefore if we move the volume of soft rubber through a space of variable curvature from a Euclidean to a non-Euclidean region it must be stressed out of its Euclidean shape in order even to enter or occupy part of the latter regions at all.

These new aspects of the situation make the mechanics of the motion of an elastic solid through a space of varying curvature even more complex than in spaces of constant non-zero curvatures. Before, it was clear that an object acquires stress when it is accelerated and maintains its stress in uniform linear motion. But in those cases, the tensor of stress was not required to change again once the object moved inertially. Now, however, the tensor of stress clearly will have to change with time as the solid moves through regions which differ in their geometry.

In this case, a simple, intuitive but quite general argument from the principle that energy is conserved tells us the kind of thing that must happen. Envisage the cube of soft rubber moving uniformly through a Euclidean region in a stress-free state. It approaches a region of non-zero curvature and moves into it. What will happen? As we saw, the stress tensor must change and the cube will acquire some energy of stress. The principle of conservation of energy in classical physics is a fundamental one. It requires that the energy of stress is gained at the expense of energy in some other form. The only candidate available in the case of inertial motion is the kinetic energy of the moving block. But it can change its kinetic energy only by changing its velocity, and momentum. So the cube will slow up, veer away from a geodesical path, begin to rotate or in some such way behave like a body acted upon by a force. However, it will not, in fact, be subject to any external force; the only forces at work are the internal elastic ones that bind its parts together.

In spaces of variable curvature, we may assume that the three laws of motion are true of independent particles moving as clouds of dust. That momentum is conserved for such particles is still a theorem. But it is not also a theorem in spaces of variable curvature for connected volumes of matter nor of course, for the individual points of such a volume. The interaction of the tensor field of stress with the vector field of momentum is so complex that we can say very little in general about how they will sum over a volume. Furthermore in certain cases, kinetic energy may be gained at the expense of stress energy, and the elastic volume will accelerate without the intervention of an outside force.

This suggests that the third law, that action and reaction are equal and opposite, is broken. This looks plausible only so long as we overlook the inadequacies of treating the mechanics of a moving elastic body by means of a single vector of momentum through the centre of mass. But clearly, the solid must change its shape and the matter that fills its volume be rearranged in order to occupy a region of space with a new geometry. This means that molecules must change their distances and orientation from one another as the solid moves. But then the intermolecular electromagnetic forces come into play, acting as the second law describes, making up a set of action–reaction pairs among the molecules as the third law requires. These intermolecular forces change the momenta of molecular points in the vector field of momentum. Just how the changes sum is complex, but the argument from the conservation of energy shows that the net effect, in general, is to change the momentum of the solid. Thus the third law is met at the micro-level in terms of action and reaction among the material points of the volume.

Clearly, space does not enter as a participant into these mechanical interchanges. What needs explanation is a mechanical effect, and there is a mechanical explanation which exhausts all the causes at work in the solid. But the shape of space still explains how these causes are brought to bear. The curvature of space does not and cannot exert pressure on the solid, as a round hole might exert *pressure* on a square peg thrust into it. Space absorbs no energy, exerts no force, enters no reaction. It plays no causal role whatever, though it very clearly plays an explanatory one. The non-causal part of the explanation of how the stress-free Euclidean cube changes its linear motion and acquires stress is simply that the space *is not there* in regions of different curvature for the matter to be disposed in a free state. That is a geometrical style of explanation, making reference to the shapes of the space in different regions.

5 Is curvature reducible?

I have a target in mind for the argument so far. It is the claim that we can discover that space has a property only if it is 'causally efficacious with respect to some events involving matter' (Hinckfuss 1975). This claim makes a reductive phase argument look plausible, for if Ψ is a mechanical property of space then it certainly seems both possible and desirable to ascribe it to a material plenum rather than to space. In the case of handedness, however, the property Ψ of space which explains how hands are as they are is neither causal nor local. In the later examples, the property Ψ is not causal, but at least it is local and may or may not vary from region to region. Is it open to us, in these cases, to ascribe the property Ψ to a material field or ether instead of to space? Hinckfuss, be it noted, does not explicitly claim that it *is* open to us, but it is worth raising the question none the less.

The quantity, Ψ, in our later example is the tensor of space curvature. I am not sure what it would mean to ascribe the curvature tensor to matter or the field, but it could hardly amount to more than saying that whatever has the quantity Ψ is *spatially disposed* in the appropriate geometrical way. Thus that an ether sea has Ψ in the way required could only be parasitic on there being a space of Ψ curvature in which the material ether could be thus disposed. If that is so, then it remains obscure to me, at least, how it *is* the material stuff and not space which has this quantity and how the geometry of its arrangement is a material property of the stuff. On the contrary, we shall certainly be obliged to deny all causal powers whatever to the material stuff if the reduced explanation is not to be more powerful than the geometric one and take us beyond the observed facts. But then, what is meant by describing as material (rather than, say, as spiritual, nugatory or null) an omnipresent, eternal, unchanging and unchanged somewhat which pervades the whole of space? What can be meant by regarding it as causally efficacious if it is strictly required to add nothing to the explanation but only to fill up the appropriately curved space?

What we can do, perhaps, is to substitute the word 'ether' for the word 'space' throughout our earlier descriptions and legislate that the result is true. But it would by no means follow that we would then have a deeper explanation, that we would have really dispensed with space or with the geometrical explanation, or that anything whatever would have got the least bit clearer in the process. We might well succeed in puzzling ourselves considerably over what the idea of the material in the description can really be and we would have run a serious risk of making it quite vacuous. The result overall would hardly seem to merit our regarding it as any kind of reduction.

Hinckfuss (1975, pp. 141-2) claims that, even where space's being Ψ is 'causally efficacious' then, if Ψ is constant, 'we would have no reason to suspect that that was so'. Any evidence for that fact would be evidence for the material causal statement:

> Distributions of matter type E always cause distributions of matter type E_1.

But what can it mean to speak of distributions of matter if not its distributions *in space*? What is a material thing itself if not a *spatial* object? Where Ψ is the affine property of curvature then, clearly, we cannot specify the E and E_1 type distributions of matter in a way which yields the material causal law without thereby stating rather directly that the regions in which it is distributed have constant curvature. This is true even if we specify only the positions of particles (up to an affine transformation). But this alone will still not give us the material causal law as a truth without further specifying each velocity vector (up to an affine transformation). If the two types of distribution are *always* to be correlated this, in turn, entails directly that the curvature is a global constant. I suspect that a similar result must hold for any pointwise definable, geometric property Ψ, whether this be curvature, the metric tensor or even some merely projective or topological concept. Which properties, other than such geometric ones, are plausibly at issue in the argument? In fact, the constant curvature of space (or whatever Ψ may be) could escape our notice only through a kind of wilful blindness to the plain import of the material causal law, since we would have the most impeccable, even irresistible reasons for perceiving the import. The only proviso to this is that the curvature might be so slight as to elude detection by our most sensitive apparatus of observation. But then the material causal law would elude detection, too.

6 Cause and spacetime

Do the preceding examples illustrate a causal, geometrical style of explanation which also holds sway in GR? That remains a complex question even if it is true that we have satisfactorily isolated and illustrated geometric explanation. Some brief remarks on this difficult matter will end this paper.

On the face of it, GR provides a very strong example of geometric explanation since not only is spacetime curvature the fundamental explanatory concept of the theory, but the idea of spacetime geometry is actually used to reduce causal explanation by gravitational force in space during time. If spacetime is flat (*i.e.* Minkowskian or pseudo-Euclidean) then a geodesic or linear path in spacetime projects onto a motion, uniform in time, along a geodesic or linear path in space. That is the case in SR. So in SR Newton's first law applies both in space and spacetime: the path of a force-free body is linear in space and uniform in time and its spacetime worldline (or trajectory, for short) is linear, too. But in GR, where we suppose that spacetime is not merely curved, but variably curved, it is no longer the case that a geodesic of spacetime can always be projected down into a geodesic of space. So while the trajectory of a force-free thing will be a geodesic of spacetime (and linear there) it may yet yield an almost arbitrary curve, including even a closed curve, when we project it into space; further, motion along it need not be uniform. That is true of the spatial path of a planet: it is a closed curve in space, but an open geodesic of spacetime. The body moves along its spatial path, not uniformly, but according roughly to Kepler's law of equal times. Given just that space and time picture of its motion we would have to regard it there as moving under a gravitational force. But we can still apply Newton's first law to it in *spacetime*, seeing its trajectory there as that of a force-free body, thus reducing gravitation to spacetime curvature, and thus cause to geometry. But we can do this only by giving up the space and time language of enduring continuants. The force reduction requires a new ontology of us. We have to take thing-trajectories as fundamental and the curvature of spacetime, which is ontologically on a par with them, as a real feature of a real four-dimensional manifold.

This conclusion was challenged by Hugh Mellor (1980) who argues that we should 'distinguish merely possible trajectories, which indeed exist as parts of spacetime, from the things which may or may not have them as actual trajectories . . . But when we talk of a thing's actual trajectory . . . [it] . . . is simply a property the thing has.' However, this distinction, were we to make it, would rob spacetime of its status as a real thing and thus rob its curvature of its capacity to explain things in the way these essays argue that it actually does. Trajectories are time-like paths in spacetime. Unoccupied, they are not mere possibilities. The ontic status of spacetime is, quite clearly, debatable. But it cannot play the role I have described for it unless unoccupied trajectories are real geometric things: time-like paths.

I venture a further brief remark on this topic. Consider SR before spacetime. The founder members of the ontology of this theory are continuant things. None of them has its properties well defined absolutely, but only relative to some frame of reference (with the exception of charge). There is some reason to prefer the proper mass, time etc., that is, the mass, time etc. relative to the object's co-moving frame. But the basis of this preference is, I submit, obscure. Why, for example, shouldn't its shape in motion be a real shape? SR without spacetime presents us with an unsatisfactory world. That is the world which Mellor wants us to accept. It is quite anomalous, metaphysically, and, clearly, the attractiveness of spacetime lies to no small degree in that all properties are possessed absolutely by four-dimensional things.

If GR were as simple as just described, then we might conclude at once, that explanation in it is a novel, powerful and more advanced kind of explanation of the style defined and illustrated in the earlier sections. But it is not so simple. The distribution of matter constrains spacetime curvature, and that sounds causal: the structure of spacetime is caused by the distribution of matter. But, as we saw in the last section, matter can be distributed only as the structure of space or spacetime permits. We cannot say either of the distribution or of the structure that one has a causal or more widely explanatory priorty over the other. Certainly they do constrain one another, but it is not yet clear that we can say more than this.

The correct way to view GR is as a field theory, and this might tempt us to claim that the individual source terms for the gravitational or curvature field are the individual mass-energies of bits of matter and not vice versa. This is largely true, but there are some subtleties in it. Significant features of spacetime remain to be fixed by choice of initial and boundary conditions. The gravitational field may be well defined for empty spacetime, not just in the case of flat Minkowski spacetime but for much less trivial structures. Nevertheless, quite certainly, one source term of curvature is the mass-energy of the body. However, this is not a characteristic constant of the body which measures something like the quantity of matter in it, as Newton thought it did. It is a function of its inner stress, too, and this is itself affected by the gravitational influence of the mass-energy distribution round it. GR is not a linear theory in which the contribution of various material objects to the total field can be simply summed (see Graves 1971, section 13, esp. p. 227).

Nevertheless this does seem to leave open the possibility of causal intervention by us to change the gravitational field. I compress a body. This changes its mass-energy however minutely, and thus alters its contribution to the gravitational field and, it would seem, to spacetime structure. So I can act upon geometrical structure even if I produce an effect rather indirectly.

This is not quite straightforward, however.[2] As earlier arguments imply, my action does not push or pull spacetime, nor does it push or pull me. But the distortion of spacetime is a store of energy in GR and this may be transmitted from point to point in the form of gravitational radiation. This may then distort another distant occupied region and thus stress a distant object. At the very least, this looks like the action of a force when we project down to space and time perspective on it. Although some of the 'forces' which arise from these projections may be pure artefacts of the projection and no more real than, say, Coriolis forces, some are about as real as can be. Perhaps, in GR, geometric explanation really can't be distinguished from causal explanation. This may mean that much of causal explanation has become geometrised. But GR surely makes spacetime something not easily distinguished from a real concrete entity with causal powers. What could be more satisfactory to ontology than that?

I will not try to settle the matter. Clearly, the situation in GR is much more subtle than it is in the cases discussed earlier, where we seem to have very clear reasons for saying that geometric structure may explain the behaviour of matter without in any way causing it. No doubt GR has changed and is changing our understanding of both the material and the spatial. I believe we can see better how this goes on if we look at GR side by side with simpler theories. The theories described in the bulk of this paper are simpler than GR in two ways: first, no basic law connects the structure of space with the density, flux etc. of mass-energy; second, space is not considered as a projection from spacetime. What I think the comparison reveals is that giv-

2 Here I have rewritten the rest of the paragraph in the light of criticisms of it in Mellor (1980) which seem to me decisive against what I wrote earlier.

ing space a role in physical explanation need not, by itself, take us any nearer to showing that space may be understood as material when we treat it as real. It is always open to us to say that spacetime is a material field. Of course, the field can be regarded as material only in a somewhat attenuated sense and there can be little doubt that field theories have changed our concepts of the material and the physical. Hence, it is by no means clear that to describe spacetime as a material field accomplishes a material understanding of space and spacetime rather than a geometrical extension of the concept of matter.

8 Is curvature intrinsic to physical space?

Wesley C. Salmon (1977b) has written a characteristically elegant and ingenious paper 'The Curvature of Physical Space'. He argues in it that the curvature of a space cannot be an intrinsic property of it. Salmon's view is that space, in respect of this and other affine structures, is amorphous. He relates this to Grünbaum's arguments (Grünbaum 1973, esp. chapters 16 and 22) that space is metrically amorphous and acknowledges parallels between the arguments which have been offered for each opinion. I wish to dispute these conclusions on philosophical grounds quite as much as on geometrical ones. Although I concentrate most on arguing for a well-defined, intrinsic affinity for physical space, the arguments extend easily to support a well-defined, intrinsic metric.

1 Coordination with real numbers

I assume that we are dealing with continuous suitably differentiable physical space unless the contrary is specified. It is worth remarking at the outset that I do not regard physical space as a *set* with points as its members but rather as a *whole* with points as its parts. It is possible to define continuity for space conceived of in this way (see Mortensen and Nerlich 1978). Physical (or part-whole) topology differs from set-theoretic topology (order type) in only minor ways: each set-interval contains a null subset but no physical interval contains a null part. I will largely gloss over these minor divergences however. Certainly, there is enough structure in physical topology to allow me to agree with Salmon on the following well-established matters. There exist countless one-to-one correspondences between the elements in any interval of the real number system and the points in any one-dimensional interval of space. I assume throughout, as Salmon does, that the intervals referred to are non-degenerate and closed. Among these correspondences, an uncountable number preserves the natural orderings of the real numbers and the points in the respective number and spatial intervals. These last correspondences *coordinate* spatial points with real numbers in countless ways. The physical topology of any spatial interval is isomorphic with the set-topology of any real number interval (save for the null set). Clearly it follows that there exist one-to-one order-preserving correspondences between the points in any one-dimensional spatial interval and the points in any other. In sum, any spatial interval can be coordinated with real numbers in countless ways.

It follows, surely, that the topology (continuity) of spatial points and intervals cannot determine their metrical properties. It follows, less simply but no less surely, as Salmon argues, that topology (continuity) does not determine affine (or projective) properties of spatial points and intervals either. Plainly, also, the coordination of a spatial with a real number interval does not tell us what its size is. It would be a little surprising if anyone thought it did tell us, since very simple and obvious considerations show that it cannot: we can measure the same interval in feet or in metres, each measure assigning a different real number, but not a different size to the interval. So far, we seem agreed on rather humdrum matters.

But neither Salmon or Grünbaum takes these ordinal, topological facts to destroy, by themselves, all hope of further intrinsic structure of space, whether it be projective, affine or metric structure. After all, the real number system itself has an intrinsic metric. For any interval in the system, this derives from properties intrinsic to the elements contained in the intervals which have measure. In particular, it derives from the intrinsic size of the numbers which are end (or limit) points of number intervals. Here, we can invoke some further property or relation of the intervals or of their constituent points such that no property-and-relation-preserving one-to-one correspon-

dence between their members exists when the new property or relation is taken into account (Salmon 1977b, p. 284).

Let us concede that points are homogeneous. All points are alike
in qualities. This needs a little hedging, however. Curvature is a
notion defined for points in a space, and it may vary from point to
point. Syntactically the curvature scalar, at least, appears as a one-
place predicate. The concession just made is not intended to grant
directly that curvature can only be a convention. Rather, curvature is
ascribed to a point because of features belonging to arbitrarily small
neighbourhoods of space round the point. It is, therefore, not a quality of the point in itself, so to speak, but only of the point as contained
in its space. Continuity is also a point-wise defined concept which
belongs to the point not in itself but as contained in its space. So is
differentiability. I shall call these *interval-dependent* properties of
points. So I concede merely that points are homogeneous in themselves, whereas real numbers are not. It follows that if the metric or
affinity is intrinsic to space, then it must be intrinsic to intervals, to
relations among points or to interval-dependent properties of points.
It cannot possibly derive from the qualities of points in themselves.
Once again, this seems perfectly congenial and unsurprising. But now
Salmon goes on to say something very uncongenial indeed:

> The absence of any reasonable suggestions as to what properties
> or relations might render two nonoverlapping segments intrinsi-
> cally equal or unequal in length lends stronger presumptive evi-
> dence to the claim that, as geometrical intervals on a line, any
> two nonoverlapping segments are isomorphic to each other, and
> that this isomorphism holds with respect to *all* of the *intrinsic*
> spatial properties and relations among the elements.
>
> (1977b, p. 284)

2 A suggestion about intrinsic spatial structures

Salmon mentions no suggestions, reasonable or otherwise, as to how
the projective, affine or metric structure of space might be conceived
to be intrinsic. Yet surely it is obvious what the suggestion has to be,
the only question being whether or not it is reasonable (in fact, the
suggestion has been made already in Demopoulos (1970), Friedman

(1972), and Glymour (1972). It is quite clear that we can get none of these structures from some complex of topological structures, since spaces which differ projectively, affinely or metrically may be topologically and differentially identical. It is equally clear that none of these richer structures can be got out of qualities which points have in themselves. We must conclude, then, that either the affinity is nothing at all or it is a *primitive* structure of space: *primitive* relations and qualities of intervals and points, including interval-dependent properties of points. I suggest, then, compendiously, that the projective, affine and metric structures of a space are well-defined (i.e. statable in an axiomatic theory) properties of it, no structure in the list being definable in terms of topology or of structures which precede it in the list.

Is this suggestion unreasonable? Neither Salmon nor Grünbaum advances any reason to reject it anywhere in his writings, so far as I am aware. Furthermore, each appears to agree in taking local topological structure, at least, as being primitive. The relation of betweenness on which the continuity of spatial intervals depends is certainly a primitive relation among the points (see Friedman 1972). More forcibly, it seems that Salmon ought to argue, in consistency, that the continuity of space is a property extrinsic to it. Elements in intervals of the real number system are intrinsically ordered because they are inhomogeneous. It is because of the intrinsic magnitude of the numbers that some are between others, and it is by means of relations derivative from magnitude that the number intervals can be ordered continuously. By contrast, points in space are homogeneous and no betweenness ordering can be based on any qualities which they have in themselves. There are countless one-to-one correspondences between points in spatial intervals and elements of real number intervals which induce different *extrinsic* orderings in the spatial points as reflections of the intrinsic ordering of the numbers. These 'alternative descriptions' are really equivalent since each one-to-one correspondence preserves all properties intrinsic to the points. Thus we have a Reichenbachian relativity of topology (continuous ordering). There are no *facts* of topology and spatial betweenness, only extrinsic conventions. Reichenbach's actual arguments for the relativity of topology (i.e. for its being conventional) were quite different from this, of course. See his (1958), sections 12 and 44.

I hope it is quite clear how we can find our way out of this wood, and equally clear that Salmon has already found it. We must insist that betweenness is a *primitive* relation *intrinsic* to the points and intervals of physical space. That we cannot define this relation in terms of the qualities of spatial points is no reason whatever for supposing there to be no such relation in fact. This seems an eminently reasonable path to take and I suggest that we may take a similar way with the further properties and relations of points and intervals required for the intrinsic projective, affine and metric structures of physical space.

An analogy with colour might be useful here. A coloured object may be red. We cannot define its redness in terms of colouredness, nor even define 'red' and 'coloured and *F*' for some predicate *F* distinct from 'red' itself. A red object may be, more determinately, scarlet, a property which cannot be defined in terms of colouredness, nor redness, nor as 'red and *F*' for some predicate *F* distinct from 'scarlet'. And so on. A thing's being red is a primitive determination of its colour and its being scarlet a further primitive determination of its being red (and of its being coloured). Just so, the affinity is a primitive determination of the topology and the metric a further primitive determination of the affinity. To be sure, matters are much more complex geometrically than chromatically. The affinity may be a conjunction of several primitive properties, but the determinate-determinable relationship is just the same for it as it is for colour.

3 The role of affine and metric language

The view which Salmon and Grünbaum state, that post-topological structures are not intrinsic to space, may sound rather less contentious than it really is. It may suggest that metric properties belong to *material spatial objects* in respect of sizes which they have intrinsically; that the linearity of its path belongs to an object in respect of its force-free nature in some intrinsic way. This suggestion is false. For Salmon and Grünbaum, the size of an object is no more intrinsic to it than the size of intervals is to space; a force-free particle has no intrinsic ability to move linearly so as to impose linearity extrinsically on its spacetime path. Their view is certainly not that the metric and affinity of space are real, objective properties of *material* things which are then projected onto space. The view is, rather, that the metric and affinity are conventional, largely non-factual determinations of scien-

tific language which do *not* reflect equally determinate structures anywhere in the world. Metric descriptions are not conventional merely
in that they are *displaced* descriptions which are conventionally
ascribed to space, though properly belonging to objects.

The theory that curvature is not intrinsic to space is thus tied to a
positive view of the role of affine and metric language, which is the
view that this language is conventional. But what does it mean to say
that these languages are conventional rather than factual, and how
does the role of these geometric conventions compare with the role
of other conventions in our theory, such as the convention of units
and the convention of systems of coordinates?

I shall try to state clearly what I take conventionalism to be.
Glymour (1972) gives a detailed explication of what it is for a property to be *intrinsic* to a set (such as an interval). The present account
of conventions is looser but perhaps more general in scope than
Glymour's, being aimed at a different, if closely related, target. A conventionalist distinguishes between *convention* and *fact.* Some sentences
in physical theories state facts, but others, though they may appear to
do this, merely reflect conventional *decisions* about what to say. A sentence may be partly factual and partly conventional, but in so far as it
is conventional, it is not factual. We can change conventions so as to
produce a new theory in the sense of a new set of conventionally
asserted sentences. But it is really still the same old theory in being
the same old set of factual sentences.

Now we can tighten up this picture of conventionalism somewhat,
without seriously extending it, as follows. A theory T, containing conventional elements has a more developed structure in its syntax than
in its semantics. We can find in L, the language of T, a proper sublanguage, *L-semantic.* Every sentence of *L*-semantic has a purely fact-stating function, and each of these sentences in T must be directly satisfied by things, states of affairs etc., if T is true. Hence there is a *semantic core* in L (and in T). Sentences in L which are not in *L*-semantic
are, to some extent or other, semantically idle. Let us say that sentences in the complement of *L*-semantic in L make up a *syntactic
periphery* of L (or of T). Sentences in the syntactic periphery of T cannot be strictly satisfied, but must be *conventionally-satisfied* if T is true.
A sentence x of L is conventionally-satisfied in a model M iff all its
proof-theoretic consequences in *L*-semantic are satisfied in M; i.e. iff
for each y in L, $x \rightarrow y$ & $y \in$ *L*-semantic $\rightarrow M$ satisfies y. Hence a set of

sentences A of L can be conventionally-satisfied in a model M which may also conventionally-satisfy a set B of L, even though $A \cup B$ is an inconsistent set. Here, A and B do not have L-semantic subsets the union of which is inconsistent. Note that M does not satisfy $A \cup B$, or any other inconsistent set since any of these have inconsistent L-semantic subsets among their proof-theoretic consequences.

These last statements about the conventional-satisfaction of sets of sentences provide a firmer basis for what appears as a prominent strategy of argument in conventionalist literature. A state of affairs is described in one set of sentences, followed by a second description syntactically inconsistent with the first. It is claimed that the second description is a *re*description of the *same* state of affairs. Without some such distinction as the one just given between satisfaction and conventional-satisfaction or semantic core and syntactic periphery it would be impossible to defend the claim. This means, and is said to mean, that only part of these descriptions states facts.

This account has some consequences for conventionalism that are worth stating. First, a condition for the adequacy of any conventionalist analysis of a theory is that it should clearly identify the semantic core, the sublanguage L-semantic of the theory's language L. Unless it is clearly identified, neither the critic nor even the conventionalist himself is in a position to know whether two syntactically inconsistent descriptions are identical in their L-semantic cores, nor just which differing conventions the two descriptions exemplify. Second, even when a candidate sublanguage has been clearly identified, the twofold case for its being the semantic core remains to be made out. It must be shown that the candidate sublanguage can exhaust the facts. It must be shown that it is not equally in question whether it is infected by conventional elements. That is, the core must be shown to have some obvious advantage over the periphery. Neither task is easily performed, and we even lack clear general guidelines which might define what needs to be done to fulfil them.

Thirdly, merely producing a pair of inconsistent descriptions which have a common core in an identifiable sublanguage provides no argument for conventionalism. For example, any metrical description of geometrical facts entails a description just in the sublanguage of affine geometry which, in turn, entails a further description just in the sublanguage of projective geometry. It is easy enough, then, to produce descriptions with the same projective core which are affinely or metrically inconsistent with each other. But so far, these are simply descriptions of different spaces which are alike in some basic (i.e. projective) ways. Alternatively, they are different and incompatible theories about the factual structure of the same space. Unless we can show that projective geometry exhausts the spatial facts (or that affine geometry does so) and is not tainted with conventions like the ones present in metric geometry, we have not begun to show anything about conventionalism in geometry. Perhaps this is obvious, but let the reader try to see how to meet the two conditions of the last paragraph as they apply to Reichenbach's argument that the number of dimensions of space is conventional (1958, section 44). It is clear that Reichenbach's strategy is to give pairs of inconsistent descriptions, but it is very obscure, at best, how the strategy is well designed to prove the required conclusion.

Quite generally, when a theory uses some convention in its language we can properly ask what function the convention performs and how it performs it. Thus, there is wide agreement that *units* are chosen by convention and that the diversity of units in practical use reflects their conventional status. What is the role of such a convention? Let us assume, for the immediate argument, that there are metrical facts. Then we need some choice of units in order to state these facts. To explicate statements about the length of intervals we need to begin, at least, with the idea of counting. Briefly, but rather crudely, we count up, for some interval, a number of equally long subintervals which partition it. A statement of length requires, then, both a number reached in a counting process and a specified item (*kind* of item) counted. Without the latter item, there is no well-formed metric sentence.

The convention, as regards units, does not lie in the preceding facts of the matter but in *the usefulness of our agreeing* to count always the same kind of item. For this reason we 'come together' to some decision, no matter which decision it is. We could say, I suppose, that nothing qualifies the metre rather than the foot to be the correct kind of item to count, nor does any thing qualify it as peculiarly apt to be assigned the number 1 in metric correspondences between lengths and real numbers. There is, indeed, no fact of such matters. It may be doubted, though, whether these remarks would shed light on any real confusions. However, this is not at all to say that statements like 'My desk is 2 metres long' are not fully factual. It is even the case that the statement 'The standard metre in Paris is 1 metre long' is factual, since it entails the invariance in length of the object referred to. Conventions of this kind do not give rise to any distinction between a semantic core and a syntactic periphery in theoretical language. While a conventional (general, customary) choice of item to count is merely convenient in metrical language, some choice of item to count is indispensable, even if we vary the choice from one occasion to another.

The case is somewhat different in other examples, such as the use of coordinate systems in differential geometry. As arguments in §8.1 suggest, we may assign coordinates for space in any of an infinite number of ways, choice among these being arbitrary. An *n*-tuple of numbers attaches to a point, in some coordination of points with numbers, in virtue of no property intrinsic to the point. Yet the reason for describing coordinates as conventions is not that there is conventional agreement among users of coordinate languages to use some one system of coordinates. On the contrary, a great deal of attention is given to procedures of transformation from any one system to any other. Presumably, the use of coordinates is regarded as conventional because it has been, perhaps always, recognised that they are features of our representation of space, rather than descriptions, true or false, of space itself. We make use of coordinates as a device to represent scalars, vectors and tensors in space, but this device, though useful, is dispensable. This recognition led, in modern differential geometry, to coordinate-free treatments of tensors generally. Hence the use of coordinates is at least *something* like the use of a sublanguage which has only a syntactic and peripheral function, but which can be (and should be) dispensed with. However, dispensing with coordinates presented itself as a problem to be solved, as a method to be developed. We could not simply abandon the use of them simply because we have recognised that they play a role only as a convention of representation.

What may be said about the function of affine or metrical languages, if they are indeed conventional in the way envisaged by Salmon and Grünbaum? I can see no function for them whatever. If the facts are comprised in topological structures, then they can be stated in the language of topology, that is, in the language of spatial properties which are invariant under homeomorphisms. We do not need metrical language to state these facts. It seems clear that, unlike the example of unit conventions, it is not at all necessary to have some metrical language in order that sentences in the sublanguage of topology may be well formed. Unlike the example of coordinate conventions, we do not need to take pains to a develop a language which is free of metric conventions. We already have such a language in current use. Yet neither Salmon, Grünbaum nor Reichenbach propose that we should somehow move to rid ourselves of these conventional, non-factual elements in our language.

In the light of these comparisons it appears, if conventionalism is correct, that no geometric language other than differential topology plays a semantic role in theory. Further, it would seem that the syntactic peripheral sublanguages of projective, affine and metric geometry really are thoroughly idle and peripheral: as conventions they perform no useful function whatever.

It might be thought that this rather sweeping conclusion is too hasty. It might be proposed that we do need metrical language in order to state quite definite empirical facts which, while they still leave the metrical language conventional, are nevertheless not comprised in topological structures. Consider two cases which might seem relevant here. Both examples were proposed in conversation by Wesley Salmon. However, as I understand him, he put forward only the first of them as an empirical fact which, while neither affine nor metrical, cannot be stated in the austere language of topology, thus providing a definite role for metrical language. I judge the second example to be no less plausible than the first.

Let us take, first, an assumption stated by Grünbaum to be empirical and factual, which he names *Riemann's Concordance Assumption*. He defines it by citing this passage from Einstein:

> All practical geometry is based upon a principle which is accessible to experience, and which we will now try to realise. We will call that which is enclosed between two boundaries, marked upon a practically-rigid body, a tract. We imagine two practically-rigid bodies, each with a tract marked out on it. These two tracts are said to be 'equal to one another' if the boundaries of the one tract can be brought to coincide permanently with the boundaries of the other. We now assume that:
>
> If two tracts are found to be equal once and anywhere, they are equal always and everywhere.
>
> Not only the practical geometry of Euclid, but also its nearest

generalisation, the practical geometry of Riemann and therewith the general theory of relativity, rest upon this assumption.

(1953, p. 192)

Grünbaum certainly does not deny that this principle provides a role for metrical language. He says the following:

> We see that the empirical truth of *RCA* plays the following role: It is a necessary condition for the *consistent* use of rigid rods in assigning lengths to space intervals that any collection of two or more initially coinciding unit solid rods of whatever chemical constitution can thereafter be used *interchangeably* everywhere in the *P*-manifold *independently of their paths of transport, unless* they are subjected to independently designatable perturbing influences. Thus, the assumption is made here that there is a concordance in the coincidence behavior of solid rods such that no inconsistency would result from the subsequent interchangeable use of initially coinciding unit rods, if they remain *unperturbed* or 'rigid' in the specified sense. (1973, p. 551)

He says similar things elsewhere about other concordance assumptions which are factual conditions necessary for conventions (for example Grünbaum 1973, p. 696). Reichenbach is distinctly more suggestive, though he cannot be taken clearly to make the kind of claim about the assumption with which I am concerned:

> This analysis reveals how definitions and empirical statements are interconnected. As explained above, it is an observational fact, formulated in an empirical statement, that two measuring rods which are shown to be equal in length by local comparison made at a certain space point will be found equal in length by local comparison at every other space point, whether they have been transported along the same or different paths. When we add to this empirical fact the definition that the rods shall be called equal in length when they are at *different places*, we do not make an inference from the observed fact; the addition constitutes an independent convention. There is, however, a certain relation between the two. The physical fact makes the convention unique, i.e. independent of the path of transportation. The statement about the uniqueness of the convention is therefore empirically verifiable and not a matter of choice.

> *. . . It is again a matter of fact that our world admits of a simple defini-*
> *tion of congruence because of the factual relations holding for the behav-*
> *ior of rigid rods; but this fact does not deprive the simple definition of its*
> *definitional character.*   (1958, pp. 16–17, italics in original)

However suggestive we might find it, the example actually provides no semantic work for affine or metrical language to do. The Riemann Concordance Assumption is purely topological in geometrical import. Though it is, indeed, a principle of invariance it assumes the invariance among objects solely of topological relations such as overlap (or non-overlap). Einstein's mention of 'practically rigid' bodies might obscure this, but Grünbaum's gloss on the passage makes it quite clear that he, at least, means merely that the bodies are solid and not subject to differential forces or 'independently designatable perturbing influences'. The geometrical content of the assumption remains topological. Hence it defines no role which cannot be fully played without the use of affine or metrical language.

My second example is a familiar one, introduced by Reichenbach (1958, section 3) and much discussed since. Consider a two-dimensional world in which it seems natural to describe the space as a plane with a hump in it. What makes it seem natural is, of course, the behaviour of our measuring rods in different parts of the space. But, Reichenbach argues, this is only because of the still greater naturalness of taking it that no universal forces operate anywhere in the space. If we assume instead that appropriate universal forces do operate in the relevant region it becomes proper to describe the plane as uniformly flat. But these descriptions are factually equivalent, he claims. We are dealing, throughout with one and the same space, the syntactic differences reflecting merely conventional, non-factual distinctions. It is this claim of equivalence which encourages the idea that the facts in this world must be exhausted by topological structures (on Reichenbach's view), since only topological properties are preserved in both descriptions. Call this world *W I*.

However, a look at another two-space world might give us pause in drawing this last conclusion. Consider a two-dimensional world in which it seems natural to describe the space (setting universal forces at zero) as a uniform plane. Now Reichenbach thinks that we can certainly redescribe this space as a plane with a hump in it (given appropriate choice of non-zero universal forces). However, despite the fact that the uniform plane is *topologically equivalent* to the humped plane, this second world (*W II*) is *not factually equivalent* to *W I*. Expanding Reichenbach's notation, although

   W I: (*G*-humped) + (*F* = 0) is factually equivalent to (*G*-flat) + (*F*≠0)

   W II: (*G*-flat) + (*F* = 0) is factually equivalent to (*G*-humped) + (*F*≠0),

neither description of *W I* is factually equivalent to either description of *W II*. So it might appear that the factual differences at issue between *W I* and *W II* are not exhausted in facts which are purely topological.

Let me agree at once that the conventionalist can and must claim that *W I* differs from *W II* in point of empirical fact. The question at issue, then, is whether the geometrical content of the conventionalist's empirical facts somehow provides work for affine or metric language or whether it leaves these sublanguages semantically unemployable, to be excised from theory without loss.

Plainly, it is the *worlds* which differ in point of fact, not their spaces, if the conventionalist account of what is factual is correct. This has to boil down to differences in the characteristic behaviour of objects in these worlds, especially of objects such as measuring rods. Naively, the difference emerges from facts like these: let each world be supplied with a large number of equal measuring rods. Then, in *W II*, we can use these rods to construct a square meshwork (or a skew parallelogram meshwork) anywhere in the space. But, in *W I*, we can do so only away from the region which we find it natural to describe as humped. This naive description certainly uses affine and metric language, but our question is whether we *need* this allegedly conventional language to state what the *conventionalist* recognizes as the facts of these matters.

The answer is that we do not need affine or metric language to state these facts. We need to replace the naive descriptions of *W I* and *W II* with more sophisticated ones before we see that this is so. In each world, let us make Riemann's Concordance Assumption about the objects we called measuring rods: none of them overlaps another anywhere, anytime. Then, in *W II*, we can perform the following rod-fitting experiment anywhere. Take four rods *a*, *b*, *c*, *d*. Fit them together so that each end point of a rod touches just one other rod only at that rod's end point. Let us say that the four rods form a cell:

they enclose a space. Take rods e, f and g and form another such cell a, e, f, g, with rod a as a base and repeat for rods c, d, e and so on. We can carry out this construction arbitrarily far anywhere in $W II$. In $W I$, however, there is a region where we cannot do so. Sooner or later the rods of the growing cell construction in this region get too crowded and will not fit together to form another cell. (In a space of negative curvature, the rods would spread too wide to touch.) These more sophisticated descriptions are couched throughout in terms of topological relations among the rods. The rods fit together in $W II$ in ways that fail in $W I$. This topological fit is invariant under the transformation (in $W II$) from uniform plane description to humped plane description. The failure to fit (in $W I$) is also invariant under the transformation from humped plane description to the uniform plane description. If there are any facts beyond those in the more sophisticated descriptions, then they are plainly violated under those transformations, and the relevant pairs of descriptions are not factually equivalent. Of course, I think that there are such facts, and that the varying descriptions of $W I$ (or $W II$) are not equivalent at all. However, the conventionalist cannot agree with this, hence he must insist that all the facts are exhausted in the more sophisticated, topological descriptions. Thus there is still no semantic work for the affine and metric languages which the conventionalist can define. Therefore he should regard these languages as eminently dispensable encumbrances on our theory.

Despite the conclusion that affine and metrical languages play no useful positive role in our theory if conventionalism is correct, it would be mistaken to suppose that they play no role at all. Further, unless the conventionalists are wrong, the role played is quite unintended and it provides us with an overwhelming reason for giving up forthwith all geometrical language beyond differential topology. The reason is that the vast bulk of people, even of physicists, are completely misled by affine and metric language if it really is conventional. Quite clearly, almost everyone outside the group of conventionalists takes metric geometries to describe structures which the world really has. Almost everyone thinks that the sun really is bigger than the earth and that light paths really are linear (autoparallel); almost all the *cognoscenti* within physics believe that the spacetime trajectories of force-free particles actually are linear (autoparallel) in a spacetime which, as a matter of objective fact, has some definite affine structure and curvature. If nearly everyone is seriously deluded about the structure of the world because of their use of this language, is it not a matter of acute urgency to *discard* the affine and metric language which deludes them? Surely this must be so unless some very tangible and weighty reason appears for retaining it as a convention. However, no clear reason of the kind appears to have been offered.

The problem why conventionalists wish to retain affine or metric language suggests a further analogy with colour. We can speak of objects just as coloured, or as red with perfect propriety; that is, without obligation to specify, there and then or for *every* purpose, which colour, which shade of red, the object has. This does not mean that we have the faintest idea what it would mean to suppose that an object might be coloured without being any particular colour. Though the idea of colouredness is more general, less committal, less rich in primitive determinations than the idea of red or of scarlet, it is, nevertheless, an idea which, in one way or another, is secondary to the idea of the specific shades of colours which particular things have. A roughly parallel observation is true of geometry.

Something like the history of geometry is encapsulated elegantly in Klein's Erlanger programme. According to this we conceive of the hierarchy of geometries from the metrical down through the increasingly general studies of the affine, projective and topological structures. These studies are framed in terms of groups of transformations and their invariants. Metrical geometry studies the properties of space and spatial geometries which are invariant under any of a certain group of transformations which can be analytically described. Affine, projective and topological geometries study spatial properties which are invariant under more and more general groups of transformations. But we should note that whichever of these groups we choose a transformation from, it operates always on a metric space. The output of a transformation from any group, like the input, is another, usually quite different, metric space. (Standardly, a metric space is said to be transformed into itself. This does not really affect the argument.) The history of our understanding of the geometry of physical space follows fairly well the direction of increasing generality of transformations. Thus it would be gratuitous (and puzzling) to suppose that, for example, topology must study a somehow non-metrical, non-affine, non-projective physical space which is *purely* topological. Certainly, the hierarchical structure of geometries does not *oblige* us to think that there can be a physical space which is simply a differentiable manifold and is no more structured than that description entails. It might well seem as mystifying to speak of topological spaces which have no metric as it would be to speak of coloured objects which have no determinate hue. Mystifying, perhaps, for the intriguing reason that we cannot visualise spaces less determinate than metrical ones. It begs a formidable question about the nature of geometries to speak of *imposing* an affine or metric structure on a topological space, which is the question 'What understanding of physical geometry have we which is not founded on the idea of transformations of one metric space to another?'

Clearly, however, non-metrical spacetimes are not unknown. It has been noticed that there is a Newtonian concept of spacetime, defined only up to affine properties, yet Newtonian gravity can be treated as the curvature of this manifold (see Misner, et al. 1973, chapter 12). However, in support of the analogy with colours, it might be observed that Newtonian spacetime would never have been taken at all seriously were it not for the prior development of fully metrical spacetime in the theories of relativity. Furthermore it was Minkowski's discovery of a metric for spacetime which led him to say that it was the real geometrical entity of which space and time by themselves were merely the shadows. We have got, here, into somewhat strange methodological waters but the voyage is, I hope, not without point.

Reichenbach says, 'Since we need a [metrical] geometry, a decision has to be made for a definition of congruence. Although we must do so, we should never forget that we deal with an arbitrary decision that is neither true nor false' (1958, p. 19). The context makes it clear that metrical spaces are intended.) The 'need' and the 'must' are neither explained nor supported.

Other conventionalists are silent on the matter. I suggest, somewhat diffidently, that the need is felt (by conventionalists as well as others) to arise from our being unable to grasp what it woukd be to see or visualise a space which literally has no metric, affine or projective structure but only a topology. Similarly the need for a specific colour vocabulary (including 'scarlet' and the like) arises from our being unable to grasp what it would be like to *see* an object which is coloured but as no specific shade of colour.

4 Some objections considered

There seem to be two kinds of objections which might be brought against a suggestion that some idea be regarded as a primitive. The first kind of objection argues that the supposed primitive is definable in terms of other primitives. Salmon himself has firmly closed the door on that approach. The second kind of objection claims that the idea introduced is somehow not intelligible as a property of whatever it is ascribed to. The door seems firmly closed against that objection, too, both because of the familiarity of metrical concepts and, *ad hominem*, because Salmon and others intend to preserve affine and metric descriptions of space and its intervals even if only as conventions.

Postulation, according to Russell, has all the advantage of theft over honest toil, and someone might think that claiming affine structure as a primitive determination of topology is very like postulation. As Ehlers et al. (1972) put it, the more complex structures fall from heaven without explanation. What has been argued already provides some sort of reply to the charge just envisaged. But, more importantly, the implied criticism mistakes the context of the present argument, which is not about the epistemology of some particular choice of affine or metric structure. It is about whether it makes any sense to regard these structures as intrinsic to physical space itself or whether the nature of physical space somehow forbids them. Thus the present attitude, that we should see projective, affine and metric structures as increasingly rich primitive determinations of deeper but more general structure, is quite consistent with any of the preferences in Grünbaum (1973), Synge (1956) or Ehlers, et al. (1972) for the use of rods and clocks, clocks alone or the combination of light rays and free fall to identify by observation particular metric, conformal or projective structures in spacetime. (See the very clear discussion of these preferences in Grünbaum 1973, pp. 730–48.) In fact, the attitude permits very clear sense to the idea that it is a factual assertion that free particles have geodesics as their spacetime trajectories. To say that the trajectory is geodesical is not at all the same (analytically) as to say that it is the path of a purely gravitating particle. Of course, none of this solves the problem of the epistemology of the factual assertion and I make no claim to be able to solve it. My complaint against views like those of Reichenbach, Grünbaum and Salmon is that they seek to solve the epistemological problem by reducing the factual content of the claim. Over and above such essentially topological facts as congruence (no overlap) of rods under transport, the coordinative definitions which render the metric make no factual claim, as these writers present the definitions. It may prove far more fruitful in the long run to retain metrical statements as being just as factual as they seem in order to work on the problem of their epistemology than it may prove to solve the problem by something uncomfortably like sleight of hand. Of course, the vanished rabbit turns up elsewhere in somewhat altered but even less tractable form. If we follow the conventionalist, we face the awkward question: why retain the factually idle language of metrical or affine geometry, since the main function it performs, as a matter of fact, is to mislead every non-conventionalist user of it as to the actual structure of space and spacetime?

5 Some constructive remarks on an intrinsic affinity

An account needs to be given of just how affine structure and affine curvature are primitive and intrinsic determinations of differential topological structure. Careful and complete accounts of this may be found in Weyl (1952), Schrödinger (1963) and in Misner et al. (1973, Part 3, chapters 9–13). The affinity is a primitive determination from point to point of what I call protovectors (directions without magnitude which do not meet the axioms for a true vector space) which can be defined in differential topology. One begins with projective structure. The primitive determination is essentially this: it is an intrinsic property of space that at any point P there is a protovector which is identical with a protovector at a neighbouring point Q. More illuminatingly, perhaps, the claim is that any physical space is such that, for neighbouring points P and Q, a protovector at P points at Q and a protovector at Q points at P and, in its opposite sense, points *from* P. These count as the same directions at P and at Q and the integral over such protovectors yields an autoparallel path or geodesic. The affinity, which fixes the curvature, then rests on a further structure. Given the protovectors at P and Q which point at one another, then any other protovector at P is identical with some unique direction (protovector) at Q. This means that we can transport directions not only along themselves, but each of them parallel to some path of transport.

In the course of explaining why he thinks that curvature is extrinsic to space, Salmon remarks very pertinently (1977b, p. 289) that the tensor calculus machinery defines the parallel vector field and the metric field by means of vectors and tensors which are 'members of the abstract vector spaces associated with the points p' of the real physical space. Focussing just on the concept of a tangent vector, there seems to be a problem for the idea that the affine connection is intrinsic in that the tangent vectors which characterise the structure are external to the real physical space characterised. Firstly, this is because the vectors are merely *tangent* to the real space and, secondly, because they lie in an *abstract* space associated with the points of the real space. In a footnote to this passage (footnote 33 of p. 302) he comments that this fact has not been explicitly mentioned in discussions about which structures are or are not intrinsic to space. I conclude with some remarks on this problem.

What is required is an account of how far the protovectors mentioned above may be regarded as *in* space and *part* of it and how they are related to the (abstract) tangent vectors in terms of which affine and metric structures are standardly defined. Tangent vectors are full vectors, conforming to the axioms for a vector space. But protovectors are to be understood as parts of paths: they are in spaces of whatever structure and cannot conform to the Euclidean structure which vector axioms enjoin on vector space. According to the methodology of Mortensen and Nerlich (1978), physical topology and physical geometry generally are not required to forgo the use of sets and set-theoretic machinery in characterising physical space. So long as space itself is not treated as a set with points as its members but as a whole with points, lines etc., as its *parts* we may take ourselves as speaking directly about the physical entity, space. So what needs to be explicated is how certain of the abstract devices of differential geometry can be understood as characterising features intrinsic to physical space without breach of this principle. A main difficulty posed by the devices of differential geometry concerns the limits of certain series. A differential *object*, which is what a protovector is, could be *part* of space only if space has infinitesimal parts of certain sorts. This is not easily done by intrinsic methods. Hence, we need to relate the protovector to an operator (the directional derivative operator) which, though not a part of space but a mathematical set-theoretic object, nevertheless characterises limitingly small parts of space. The operator *characterises* what is intrinsic to the space while not being itself a spatial object. Let us consider what this might mean.

In a suitably differentiable physical space, S, consider all the paths through a given point p which have a common tangent at p. Call these paths, including the tangent itself, a co-tangential set. Any diffeomorphism on S will map these paths into another co-tangential set. Generally, if a and b are paths in S which are co-tangential at p, then the image curves $t(a)$ and $t(b)$ are co-tangential at $t(p)$ where $t(x)$ is a diffeomorphism of S. Let each path in a co-tangential set at p be parameterised by a map from the interval $(0, 1)$ $(0 < p' < 1;$ p' the coordinate of $p)$. Consider, for each path in the set, the decreasing set of spatial intervals generated by the real number intervals (a,b) $(0 < a < p' < b < 1)$ in which $|a-b|$ decreases without limit, the $n + 1^{th}$ interval in any series being a proper part of the nth. Consider the nth terms of all these series. Taken together, they give us an array of small parts of the co-tangential paths and, intuitively, these small paths approach each other arbitrarily closely as n increases without limit. (A more rigorous account is given in Misner et al. 1973, pp. 226–30. See also Coleman and Korté 1982, 1993). It is this feature of co-tangential paths (parts of the space) which constitutes the protovector and which the directional derivative characterises. Thus while a tangent vector is really abstract, it characterises a feature which is intrinsic to the space.

However, we still need some account of what the tangent vector is and of how it is used if we are to reassure ourselves that its use in defining the affine connection does not forbid our regarding the connection as intrinsic to the space.

Our target is an n-dimensional physical space S and neighbouring points in it, p and q, which are to be affinely connected. Let E be an *abstract* Euclidean space (a *set*) of $m + n$ dimensions, sufficient to permit the following constructions on an n-dimensional differentiable neighbourhood. Let S' be an n-dimensional abstract subspace of E which is isomorphic to S in the neighbourhood of p and q. S' will be embedded in E with some exterior curvature which need not concern us. We may construct in E an abstract space tangent to S' at p. In this tangent space are the tangent vectors which are used to fix the affine connection from point to point. Clearly, in S', there are abstract objects, which are the images of the co-tangential paths through p in S. So we can construct image protovectors in S'. Roughly, a limiting operation in the $m + n$ space E enables us to relate the external Euclidean tangent vectors one to one with the protovectors of the (abstract) embedded space S' and to claim that, in the limit, they are 'arbitrarily close' to one another in E. Thus the tangent vectors of the abstract space which are tangent in E to the (abstract) S' will also permit us to characterise the set of series of *physical* parts of our *physical* space S. Though these vectors tangent to S' in E are abstract things, not parts of material space, the isomorphisms between the abstract S' and S permit the one external object to describe the other, intrinsic, one. It is required simply that our target physical space, as a structural object, identifies a set theoretic space which is isomorphic with the abstract n space S', embedded in E. The advantages of this treatment are clear. The abstract space gives us true Euclidean vectors which are ideally simple, conceptually, and ideally easily related to one another at neighbouring points. Thus the affine connection is a relation between neighbouring points in physical space, the connection characterising certain co-tangential paths at p and at q together with all the protovectors at p and q as connected by a path between them. The connection is intrinsic and physical, then, despite the abstract and external devices which are used to characterise it.

To sum up: I agree entirely with Salmon and others that we cannot define a metric or an affinity for a space in terms of its topology. Nor can we base them on properties intrinsic to different spatial points, which are homeogeneous. However, I see every advantage and no objection to claiming that affine and metric structure are primitive, intrinsic determinations of space.

9 Holes in the hole argument

1 Introduction

John Earman and John Norton claim that modern spacetime realists (substantivalists) face a new problem: a realist can't also be a determinist. They argue this both separately (Earman 1989; Norton 1987) and together, notably, in Earman and Norton (1987). The problem has been tackled here and there, mainly in attempts to find a picture of determinism which evades the problem. (Butterfield 1987; 1989; Maudlin 1990a).

This has my sympathy, but I look for another kind of reply. I explore the analogy (which Earman and Norton draw) with Leibniz's classic objection to Newton's absolute space. There are at least two ways of understanding both the analogy and the hole argument. One way, which I'll call the metaphysical way, stays fairly close to the historical Leibniz. Its theme is whether realism or relationism in regard to space (spacetime) is ontologically proper (legitimate, necessary etc.). Another way, a more modern one which I'll call the extensional way, simply sees the issue as a cost–benefit calculation in theory choice, pricing, as it were, one set of metaphysically innocent theoretical entities against another. I argue that, whichever way we interpret the argument, we find no strong reason to abandon realism. Understood in the metaphysical way, Leibniz Equivalence simply fails to establish what is claimed for it. Understood in the extensional way, the indeterminism claimed by the hole argument is something a realist cannot reasonably care about. In any event, the indeterminism is chosen within a realist framework for plain, straightforward reasons of physics.

First, a sketch of the metaphysical way: Leibniz saw 'Leibniz Equivalence' as more than a mere *motive* for treating space as a chimera, as an ideal thing. He thought he could *justify* doing so – legitimise, even enjoin, taking space as a device of representation, a merely ideal construct embedding real spatial relations among things. Leibniz aimed to make us *understand* this way of viewing space. (In fact, I think Leibniz wanted more, but I won't pursue that now.) A motive to dismiss space from your ontology is useless without a legitimate way of dismissing it. So Leibniz tried to show not just that Newton's physics is indeterminist, but that it is detachable from any space (spacetime) setting. Specifically, he argued that the spatial relations which things have to things are not dependent on (and may be detached from) the spatial relations which things have to space. Nothing less can legitimise regarding space as a representation. The hole argument needs an equally strong result, if we are to pay it metaphysically. In fact, Leibniz failed to establish detachability, though he thought he had done so (see §9.6). He mistook the significant conservative role of the symmetries of Euclidean space for something empty, not a geometrically explained invariance but a metaphysical extravagance. He established only a weak kind of indeterminacy. The hole argument fails to establish detachability, too, and for the same reason. It mistakes the significant conservative role of manifold symmetries for a triviality. I hope to show this by cutting 'surgical' holes in the manifold to destroy the symmetries and cripple the argument. (Surgical holes are not holes in the weak sense of the hole argument; they are the literal removal of points or open balls from the manifold.) If I'm right, surgical holes in the manifold are also holes in the metaphysical hole argument.

My sketch of the extensional argument is based on Mundy's (1983) and Friedman's (1983, chapter VI) elegant and incisive accounts of relationist representationalism. On this interpretation, there is no metaphysical issue whether a realist can properly posit spacetime in the ontology of his theory. Of course he can. Nor is there a metaphysical question whether the relationist can properly treat space (spacetime) as a representation embedding the privileged (material? observational?) entities and relations of the theory. The relationist no longer needs to justify his representational account. There is no metaphysical difficulty in the idea of the restriction of a model (and its properties and relations) to a subdomain. The distinction between a restriction's being an embedding and a submodel is clear. There can be no problem of understanding here so long as we treat models as sets (and how else are we to treat them?). The realist wants to regard the restriction of the standard full model to the physical objects (occupied spacetime points), their properties and relations, as a submodel of the full model; the representational relationists wants to see it as embedded. Here, there can be no question how to *understand* the restriction of the model to the objects, their properties and relations among them: the domain of the restriction and the sets that constitute the restricted properties and relations are simply subsets of the domain and the counterpart sets in the full model.

I confess to old-fashioned reservations about the adequacy of the extensional argument to address all our concerns. I suspect that I am by no means alone in this. While I am convinced that the credentials of space or spacetime to form part of the ontology of theories in physics have been established as impeccable, I concede that others may reasonably be unconvinced. The relation of space to perception and action is too subtle and strange, too rich in problems about how we can formulate and generate ideologies for science or for other cultural enterprises. By contrast, I am quite unconvinced that the embedding/submodel distinction can address some real problems: what does the relationist *mean* by saying that he believes only in the things, properties and relations of the restricted model? How can the *concept* of space simply drop out of his ontology as a mere picture? How does the relationist construe spatial relations *in intension?* What, for example, can it mean to say that *x* is at a distance from *y* unless that meaning includes there being a path joining them along which distance may be defined? Is it not part of what it means for a relation to be spatial that it be mediated by what is between the related things? (See Nerlich 1994, chapter 1.) Appeals to the subset relations and the embedding/submodel distinction afford no help with these questions. Arguments supporting a resolve to countenance no question of meaning or modality as proper in philosophy are neither unknown nor undistinguished, of course. Perhaps the line I am taking here has something in common with Earman's complaints about modern relationist theories in his (1989), chapter 8 (see e.g. p. 166). At least his remarks there make me forbear to ascribe to him the extreme representationalist interpretation of relativism that I call the extensional one.

I have no space to debate these questions of method here, nor the general merits of the representational relationist's account. So, in what follows, I consider both the metaphysical and the extensional interpretation of the problem; each is of interest. But if the issue really is extensional and quite purged of metaphysical pain, is it not best to leave the problem how best to formulate the theory to the mathematicians and physicists who have to use it? It seems unlikely that philosophers, mere musers upon theories, will have a better view of what's the most deft way to formulate them. The users' choice is already clear: they retain the manifold.

(Let me glance briefly at the question of confirmation, explanation and ontology. Here I follow Whewell (1840, vol. II, pp. 212–59) and Friedman (1983, VII, §3): in the ontology of science one tries to posit things which yield unities in explanation and thus in confirmation. So if positing space as a real thing allows such a unity we are right to posit it. However, the idea of such unity can't itself be fully understood extensionally, I believe: unity includes oneness in ideology, in meanings.)

Earman and Norton claim, plausibly, that we should not abandon determinism unless it is for 'reasons of physics'. This is a claim whose force is best understood within the extensional way of interpreting the question. I argue that, within a realist perspective, the hole argument gives us precisely reasons of physics for abandoning determinism. To this end, I look briefly at the strong analogies between Newtonian indeterminism in respect of 'infinitely fast particles', the indeterminacy of Leibniz Equivalence symmetries in Newtonian spacetime, and the symmetries which are the basis of hole diffeomorphisms.

2 The hole argument

But, now, what is the hole argument? There are several good accounts of it. Mine owes much to Butterfield (1989).

Our physical theories, from Newton through general relativity, can be written as spacetime theories and thus neatly compared. Any model of such a theory is an n+1-tuple $<M, O_1, O_2, \ldots O_n>$ where M is a manifold and the O_i are geometric objects defined on each point as differential geometry requires. We can compare different models of a theory by means of diffeomorphisms from the manifold M of one model to M', the manifold of another. (Of course, the manifolds must be like enough to allow a diffeomorphism, a point that will be seen to matter.) Any diffeomorphism d is a smooth 1–1 map from M to M'. It induces a 'drag along' map $d*$ which carries the O_i into O_i* on M', which can then be compared directly with the O_i' there; d induces the map $d*$ on vectors, tensors, projecive and affine structures etc. If the O_i* agree with the O_i' within some region of M', then the models are, so far, alike and if not, not. We can use diffeomorphisms to map M onto itself, which gives us the most vivid picture of what it is for the theory to be deterministic.

Now suppose that our theory is like this. The manifolds of its models typically contain regions of some type S (time-slices, for instance). Consider any diffeomorphism d (with its associated dragging function $d*$) which maps M onto itself and maps some S region to an image S region where the O_i* are identical with the O_i there. The question is whether d, $d*$ induce an identity of the O_i* with the O_i *everywhere* on M. Now, if every d which maps S-type regions in the way envisaged does result in such an identity throughout M, then the theory is S-deterministic. By contrast, let **d** be a diffeomorphism (with associated dragging function **d***) which maps a region of type S in M to another region of the same type (or, more vividly, maps the S-region onto itself) so that all the O_i* are identical with the O_i in the image S-region. Suppose, however, that the O_i* are not identical with the O_i elsewhere in M; the theory fails to be S-deterministic.

Other accounts of determinism are debated (Butterfield 1987; 1989; Maudlin 1990a). Whether the argument establishes indeterminism in the most useful sense is beside my main point, so I pursue it no further.

The hole argument aims to show that spacetime theories of the kind we are focussing on all fail to be S-deterministic for an important type of manifold region. Let an *empty N-hole* be a neighbourhood of the manifold within which the matter tensor is zero; **T=0**. The hole argument applies to any such empty N-hole; it is possible to find a diffeomorphism such that both it and its dragging map differ from the identity map only within the *N*-hole and smoothly with it at the boundary. So there are arbitrarily small empty regions of any space-time within which the fields may vary their manifold locations without perturbing any other relations either within or beyond the hole. Clearly such theories violate *S*-determinism as just described. Regions of the relevant *S*-type consist of the complement in *M* of an empty *N*-hole.

This indeterminism will be a matter of fact, if the manifold is a real thing and not merely a representation. So, if we are realists (substantivalists) about the manifold, indeterminism is inescapable. But now we have accepted indeterminism with no reason of physics to recommend it. We should never do that. So we face a dilemma: we can properly hold to determinism or to realism (substantivalism) but not to both.

That is the hole argument in a form that admits either a metaphysical or an extensional gloss.

3 What is substantivalism?

For spacetime theories in which the geometric structure and appropriate matter-field equations constrain one another (at its simplest, where we can equate the **G** tensor with the **T** tensor), it has become usual to identify just the manifold as spacetime itself. General relativity is just such a theory. I don't need to oppose that view of spacetime directly, though to grumble about it has a relevance to my case (which does not depend on the grumble, however). General relativity has seen a collapse of the view that physics is about matter in motion in favour of a view that it is about fields; they are not reducible to (nor intelligible as) emanations of corpuscular matter as the nineteenth century conceived of it. Of course fields are physical, but physics has been geometrised quite as much as geometry has been made material. The new concepts have left much (not all) of the old absolutism/relationism oppositions behind. Spacetime has become physical and, indeed, a substance in general relativity – or so it seems best to say.

But that is not the way in which the literature views the matter, as Earman and Norton make clear. That, in general relativity, the metric tensor constrains the matter tensor and is constrained by it would seem to fulfil amply the most soaring ambitions of any spacetime substantivalist. Yet just this fulfilment seems to be invoked as a refutation of substantivalism at the expense of relationism (though what relations have specially to do with it is less than clear). But what else would one want to describe as substantivalist other than the view that spacetime stores energy and mass, that particles may be generated out of and decay into spacetime structure? Certainly not the view that *is* described as substantivalist – that manifold structures, despite being undetermined by the matter tensor, are real ones. A manifold, bare of other structure is, just for that reason, not a substance. Still, the manifold structures of spacetimes are, arguably, none the less real. Whoever argues that they are real is best described as a structural realist in that respect, and therefore not a substantivalist. Else he is described as claiming somewhat more than he undertakes to defend.

Now it is argued by Earman and Norton that identifying spacetime with the metrical spacetime of general relativity will blur a distinction between container and contained which plays a crucial role in the historical debate. Perhaps so, but only because general relativity inevitably blurs that distinction. I find, in turn, the distinction between manifold spacetime and metrical spacetime dubious as a distinction *among entities* in the way this seems to require. The metric, the connection (and so on) are not really objects contained in a manifold: they are properties of spacetime just as the smoothness and the local Euclidean topology (which are manifold properties) are properties of spacetime. Here again, an extensional approach may count that as of no substance, but this, again, serves to convince some of us that such an approach cannot address legitimate concerns. If we can only retain the idea of space as a container by passing off its structures as objects contained in it, then it looks better to sacrifice containment.

The position defended in this paper, then, is realism about spacetime as part of the ontology of spacetime theories generally; it is substantivalist about such structures as the metric, the affinity and the like, realist but not substantivalist about manifold properties. I have tried to write the paper so that it works independently of the stand taken in this section. But it would be idle to pretend that I think the stand irrelevant.

4 The Leibniz analogy

The hole argument is somehow like Leibniz's classical objections to Newton's realism about space. It appeals to Leibniz Equivalence. Earman and Norton draw the parallel in some detail; Maudlin and Butterfield both endorse it. In the Third Paper, §5 of his correspondence with Clarke (Alexander 1956), Leibniz argued that we can interchange all matter east to west or transport all of it the same distance in the same direction without changing any of the spatial relations among things. Uniform motion (and, indeed, acceleration or rotation) of every object at the same speed in the same direction leaves all thing–thing spatial relations unchanged (Fifth Paper, §52). Doubling the size of everything simultaneously also changes thing-to-space relations while leaving the sum of thing-to-thing relations unchanged. (This last is Poincaré rather than Leibniz.)

Clearly these objections are all of a piece. They give us at least an indeterminism in respect of thing-to-space relations; they somehow invite us to regard space, the entity which is the would-be bearer of the indeterministic relations, as a pseudo entity. To see exactly how, we need a sharper focus on what Leibniz himself thought he was doing in this argument. He aimed to 'confute the fancy' of those who posit the chimera, space, as a real thing: to show space as merely ideal, merely a representation. He wanted a *justification* for dismissing space as a representation. Thus he aimed to show that thing–thing spatial relations are always detachable from thing–space spatial relations. He wanted to conclude that differences in thing–space spatial relations can never make thing–thing spatial relations differ; that thing–space relations are in principle indiscernible. He needs the thesis that they *cannot* make those differences, else the difference they can make can't spring from a mere representation. Only given this conclusion could he claim that *all* space can be doing in such an account is representing. The prevalent modals in this account qualify it as metaphysical. It defends a far stronger thesis than that determinism fails for thing–space relations. Leibniz offers us a legitimised choice in that he shows (tries to show) how the alternative to realism (substantivalism) can be thought through. A proof of indeterminism may make us *wish* we could do without space, but fails to show how we *can* choose to abandon it.

The interpretation of Leibniz's detachment argument is not without its problems. Leibniz was not a relationist in the modern sense. I understand him as construing space, time *and spatiotemporal relations* as ideal or phenomenal, though well founded on the nature of the monads which make up the real world. But Leibniz's rejection of a vacuum suggests that he retained some surrogate form of mediation even for these ideal spatial relations. In his careful and scholarly study (1979), with its interesting title 'Was Leibniz a relationist?' Earman makes rather similar observations. As I understand him (see esp. p. 268), Earman sees the detachment argument as providing us with no more than a motive for writing space out of the ontology of physical theory. I think this is correct, but interpret Leibniz somewhat differently. I believe that it was his intention to show how the spatial relations of things to things could be detached as logically independent from the relations of things to space. I take it that this is a somewhat stronger interpretation than Earman offers in his (1979). It is very much this approach to the Hole Argument which Earman takes in his (1989).

Now perhaps Earman and Norton (1987) are not arguing anything very like this, though neither do they appear to reject it. They acknowledge no need to justify representationalism. Nevertheless, an argument for its justification is open to us in respect of manifold properties for the spacetime theories specified. Further, if it can be got to work, it is the most powerful form of argument in the field. We can vary field-to-manifold relations freely inside the N-hole without varying field-to-field relations there at all. To justify representionalism in this case, we would need to show how the field-to-field spatial relations can always be detached as a whole system from field-to-manifold spatial relations, within these arbitrarily small empty regions. Then we can abandon field-to-manifold relations as mere pictures without damage to field relations; so the manifold (spacetime itself, as fashion has it) *can* drop out in the same movement. If we thereby save determinism, we also *want* to drop it. But metaphysically, indeterminism functions merely as a motive; there is no suggestion how it may function as justification.

5 The failure of Leibniz's metaphysical argument

Leibniz's argument for necessary detachment is invalid; but if only detachability can justify the distinction between space as something real and as a mere representation then he failed to justify it. (Of course, we still have the bare claim of the extensional argument: that representation is distinguished from reality by way of a set-theoretic embedding/submodel distinction.) Leibniz could not have seen that these challenges to Newton depend on *attachment* to geometrically specific kinds of space, most notably to the Euclidean kind. He had no reason to doubt that the symmetries his argument exploits were necessary features of a metaphysical type: he was wrong.

(*a*) Leibniz's mappings of the system of thing–thing spatial relations onto new thing–space spatial relations by uniform motion or acceleration are invariant only if the space has no constant non-zero curvature. They depend on the existence of parallels along which all the particles can move in the same direction.

(*b*) So does the Poincaré doubling transformation. That depends on there being similar figures of different size. Among geometries of constant curvature, these features are unique to Euclidean space.

(*c*) If we interchange east with west, the transformation will vary in respect of thing–thing spatial relations if the curvature is variable.

(*d*) The same is true for displacements three feet east: each object needs to be mapped to a region of curvature appropriate to its shape.

Thus detachability fails: thing-to-space differences can force thing-to-thing changes. They are not indiscernible. They are not of a metaphysical type which dooms them to be mere representations.

Leibniz mistook the *geometric* features of special types of space, especially symmetries, for the *metaphysical* features of spaces quite generally. A glance at spaces with other geometries highlights this fallacy immediately. In particular, he mistook the significant geometric symmetries of classical spacetime, $E_3 \times R$, for general, necessary, metaphysical symmetries. Seeing these as trivial, he seeks to detach and repudiate them. But Leibniz needs just what he wants to repudiate – an attachment to a symmetrically structured space. Everything he has to say to Newton hangs by entailment from particular geometric structures of Euclidean space (or at least space of constant curvature). None of it has anything to do with the ontic type of space. Thus the metaphysical case for detachment, though not for indeterminism, fails.

The aim of this argument is to make it clear how we can be justified in writing space out of physics as a representation. Unhappily for Leibniz, he showed only that a physics in Euclidean space is indeterministic in a special sense. The case for detachment is false, as recourse to non-Euclidean geometries reveals at once. Differences that conserve are not at all the same as 'differences without a difference'. Not only that, it reveals how thing-to-space differences may enforce thing-to-thing spatial differences, which shows, in turn, that space as a metaphysical type is not condemned to a representational role. It shows that it is something at least very like a concrete particular with causal powers. So far from proving that we *must* jettison absolute space, Leibniz failed to show even how we can do so. He says nothing that can license dropping space from our ontology. His argument is powerless to justify any reductive procedure, though it has been widely thought to do so.

6 Reasons of physics?

Does indeterminism provide us with a strong motive for wanting to abandon absolute space?

First, some remarks on physics and detachability. Newton's mechanics is tied tight to Euclid's space. Suppose we apply Newton's laws to point masses in motion in some space of positive constant curvature; it is not hard to see that the conservation of momentum must be an early casualty. The path of the centre of mass of two particles may not even be continuous, in appropriate conditions (as pointed out to me by P. Catton and G. Solomon[1] in correspondence). On one plausible procedure for projecting a quasi-Newtonian physics into the motion of extended bodies through such a space, the body will move under stresses analogous to those experienced by an extended body that rotates in Euclidean space (§9.7). So changing the space to

1 Their example works like this. Suppose two equally massive particles begin at zero degrees latitude and longitude, **a** moving directly north and **b** moving east along the equator at half the speed of **a** (both constant speeds). The path of their centre of mass **C** is not a geodesic, nor is it traversed at constant speed; it is directed roughly north-north-east at first, but tips gradually southward, intersecting the equator again when **a** itself does; it winds on down to the south pole when **a** reaches the equator again, whereupon it immediately disappears and re-emerges at the north pole, continuing in a reflection of its earlier motion.

which our physics is attached may fundamentally change the rest of physics. This point goes beyond the failure of Leibniz's invariance claims. For a realist, it ties the structure of space directly into physics. The hope of detaching thing–thing spatial relations from thing–space spatial relations in physics is delusive.

Newtonian physics is indeterministic *in a weak sense* as Leibniz clearly showed. Nothing determines whether or not everything is moving uniformly through absolute space, though the total disposition of matter at one time does determine it at all others, so that it is *S*-deterministic (with reservations to be raised shortly). There are isometries which allow us to move space over itself and drag the 'matter fields' (all the objects and their properties) with them. What should we say about this independently of what's already been said about the metaphysical argument? There are two ways to go here.

One is to rewrite physics as a spacetime theory. We need no longer think that acceleration – 'change of motion' – entails that there is motion. We can invoke the 4-connection instead of rest and motion. Acceleration becomes worldline curvature, so it's a property intrinsic to trajectories. The gain is parsimony; we save ontic expense. Spacetime is no less prodigal in ontic structure than space and time. It has a connection and a metric quite as much as Euclid has. But it does allow us to omit a geometrical structure – we need no vector field to rig spacetime so as to define an absolute frame. No such move was open to Newton, nor to Leibniz, though both he and Huygens groped for it. Given their conceptual horizons, Newton had the best of the argument. In turning to spacetime to rid us of these indeterminisms are we calling on reasons of physics?

Yes, provided that the upshot of Minkowski's work is not a mere rewriting of Einstein's 1905 special relativity. I argue that it is not (§4.3). Spacetime provides a background against which we can speculate intelligibly that, for instance, there are superluminal velocities (tachyons), that the photon could be massive and thus not occupy the null cones; i.e. that spacetime structure, not photons or other 'causal' elements, identify the cones – all this consistently with special relativity and inside its framework. However, this move lies tangent to the line I want to defend, so I leave it.

A second way to go would be to give some constant curvature to space in a classical setting. At one stroke, this would give us determinism at the expense of Galilean relativity. (It would still leave some

Leibniz symmetries afloat, but ignore these for now.) More importantly, it has the usual advantages of realism; it is clear, conservative and familiar in its ontology. It would impose a unique frame of reference, in principle, if not necessarily in observational practice. Further, though we make no ontic savings in structure this way, neither do we spend more than we would on familiar Euclidean space. That its curvature is everywhere zero does not mean that there is no connection in Euclid's space, just that there is a simple one. Suppose this option had fallen within the conceptual horizons of Newton and Leibniz; further suppose Leibniz had argued against Newton that we should take this second way to save determinism rather than to regard space as a chimera.

Newton would have won this imaginary debate since his reply to Leibniz would be devastating! There is no evidence at all for any curvature; flat space yields a much simpler theory. Constant curvature might rescue determinism but would thrust upon mechanics obscure principles of metaphysics (Identity of Indiscernibles) which have no place in natural science. I suggest that it is clear that the decision would be made Newton's way; more to the point, in this context of realism, the reasons for making it, and thus for setting determinism aside, would be reasons of physics.

Newton's theory is indeterministic, quite differently, in a strong, S-deterministic, sense. It admits arbitrarily fast particles or other causal influences (Earman 1986). Having no reason to postulate any particular finite maximum speed, Newton could have fastened on one only at random (though, of course, it would have to be greater than any observed speed). Nor had he reason to postulate a non-zero curvature; he could have fastened on one only at random (but the curvature would have to be too slight to have been already observed). It is not clear on what principle the reasons differ in the two cases. Good reasons of physics rule out an arbitrarily attained determinism in much the same way in both cases.

Nevertheless there is a marked difference between the two cases of indeterminism. The respect in which the theory is deterministic will have much to do with how much we should worry about it. If no finite limit can be placed on particle speeds, then something can race in 'from infinity' and change the predicted physics in respect of privileged physical properties (on anyone's view of privilege). We can't, for instance, predict the spatial relations between a local system of point particles if they may be disrupted in this way. Worrying indeed! But not when the indeterminism is the result of geometric symmetries. No privileged properties are undetermined in that way. I am arguing that the distress caused by the infinite speed case lies *not in its being a case of S-indeterminism* rather than weak symmetry-indeterminism, but in the nature of *what it is we cannot determine.*

If these arguments are good, then realists – including actual practising scientists – sometimes won't adopt deterministic versions of theories even when they know where to find them. Even when the indeterminism is both *S*-indeterminism and also otherwise a worry! No classical physcist ever postulated a finite maximum speed for causal influences. Physical reasons may be decisively against determinism.

Of course there is an awkwardness about making sense of distinctions which we do not care about: we can allow Leibniz that point. Euclidean space allows symmetries of motion which mean that there are differences in physics *about which we have no reason to care*; about which, indeed, we *cannot* reasonably care. That is simply an observation on what symmetry is. They don't and can't intrude on our expectations in regard to privileged properties. For us, but not for Leibniz, the symmetries are direct consequences of our best choice of realistic postulate, and they should come as no surprise. That we cannot sensibly care about the indeterminism has nothing to do with a metaphysical peculiarity of space. It springs from our choosing a symmetrical geometry – a best-evidenced choice among a range of realistic options. It's a bit awkward that there are differences which our postulate makes so clearly pointless; yet we bow to the fact that there is evidence for it, but none for deterministic postulates. Within our realistic range of postulates we defer to reasons of physics.

7 Consequences for the hole argument

Granted this picture of the role of Leibniz's Equivalence in the classical debate, how relevant is it to the newer-fangled objection?

Consider first a metaphysical interpretation of the hole argument. Like Leibniz's argument, it can't get off the ground unless realism postulates the right kind of thing to begin with. It needs manifolds which will permit the relevant diffeomorphisms. That draws on symmetries of the usual manifold. These don't characterise a meta-physical type; they are not necessary features of manifolds. The usual manifold doesn't have much structure, but it does have some. It is everywhere locally Euclidean in its topology and everywhere smooth. That ensures that we usually postulate a highly symmetrical manifold. But we aren't obliged to do it by any metaphysical priniciple, and sometimes we don't.

If a manifold has surgical holes in it then it will lack the local symmetries without which the hole argument can't begin. (Cusps would do just as well. Surgical holes are not holes in the weak sense of empty N-holes; they are the literal removal of points or open balls from the manifold.) The local N-hole symmetries to which the hole argument appeals must be *necessary* features of manifolds if they are to fix an ontic type, to float detachability, not mere indeterminism. But if we cut surgical holes inside N-holes, the symmetries will go with them. Yet a manifold with surgical holes is still a manifold; given appropriate field structures, it can model a spacetime theory. That gives us a way to mount criticisms parallel to those we urged against Leibniz. The hole argument can't start unless it picks the right kind of manifold. But then it is too shallow to float detachability; it can't legitimise dropping the manifold as a mere representation. It needs, but fails, to show that manifold differences can make no field differences. It mistakes a significant conservative role which rests on a geometric, not an ontic, type for a triviality.

I am not claiming that a relationist can have nothing to say about surgical holes (see Earman 1989, chapter 8, §6 for some observations on this). The relationist's strategy here is to defy the realist to show how s/he can put manifold differences to some discriminatory use. Surgical holes illustrate that use. So do cusps in the manifold.

If we want to legitimise viewing the manifold as a mere representation we need detachability. But the hole argument gains only indeterminism. It cannot show that manifold differences are impotent to make field differences, for they plainly do make differences. First, any S-hole in the N-hole will limit the movement of the fields over it. The S-hole will get in the way. It will not block every diffeomorphism, for we can drag round it, but it will block some since we can't drag across it. For every hole diffeomorphism from M to M' we can construct another manifold M'' with a surgical hole in its N-hole which will block the counterpart of the first diffeomorphism. (This description is a bit loose, but I take it that the message is clear enough.) That sinks the claim that manifold differences make no difference. We can even surgically remove all of the *N*-hole. That will prevent us sliding the fields about within the hole and do it by purely manifold differences.

We need to be clear about what this revamped objection does. Unlike the objection from curvature in the case of Leibniz's original argument, the objection from surgical holes does nothing to restore determinism. But determinism is a red herring in the metaphysical argument. If relationism wants to justify representationalism, then it needs detachability as a metaphysical character of manifolds. Leibniz's argument does not provide it.

I go on in the last section to reflect on surgical holes in the extensional interpretation of the hole argument.

8 More reasons of physics

Clearly, surgical holes in *N*-holes will limit the diffeomorphisms there; but more, they make an independent difference to the physics itself in familiar ways. Singularities in spacetime are very much a part of general relativity; physically realistic singularities in spacetime occur within matter fields in familiar ways. Surgical holes in empty regions play a part in the formal development of the idea of a spacetime, where we are less concerned with the physical realism of models (see, for instance Hawking and Ellis (1973, chapter 6). We know what cutting *S*-holes does; we have sound evidence not to postulate empty *N*-holes with surgical removals inside them. Our best choice of *physically realistic* manifolds postulates them with the symmetries which the hole argument exploits. The symmetries make for differences which realists cannot sensibly care about, despite *S*-indeterminism. Yet if determinism fails because of manifold symmetries, it does so for reasons of physics. Change the manifold and you change physics.

What now of determinism as a motive for abandoning realism in respect of manifold properties (taking our question to fall within the extensional argument)? Grant, for this argument's sake, that since many diffeomorphisms may remain open for every intact *N*-hole, *S*-indeterminism is established. For realism, it is now perfectly clear that the diffeomorphisms depend on symmetries in the manifold. Despite

S-indeterminism, the result is not a worry. Every relevant diffeomorphism is, in fact, an isometry.[2] As before, the symmetries mean that
there are differences *about which we have no reason to care*; about which,
indeed, we *cannot* rationally care. That we can't sensibly care does not
rest on necessary features of manifolds which make them metaphysically non-physical or unreal. It has nothing to do with ontic type. It
rests on the real manifold symmetries of the spacetimes which we
have best reasons of physics to choose. Indeterminism, whether weak
or S-type, gives realism no good reason of physics for concern about
the manifold structure it posits.

We can now see clearly what the role of manifold symmetries is in
spacetime theories. They provide for the well-evidenced, smoothly differing locations for field quantities. Neither Leibniz Equivalence nor
the hole argument even appear to give an indication how else we are
to provide for them. (That is simply the practical side of the conclusion reached earlier that Leibniz's metaphysical tactics are powerless
to legitimate our dropping the manifold from the ontology of spacetime theories.)

Granted all this, what is it really best to do? Earman(1979; 1989)
seems to admit the need to find another way to provide for smoothly
differing locations – by means of an abstract Einstein algebra as surrogate for the manifold. But it is unclear what this has to do with the
hole-argument. The tactic was first introduced into the literature by
Geroch (1972) in order to find a non-punctile base for a quantized
relativity – which is irrelevant to the hole argument. If such an alternative basis is needed before the manifold can be jettisoned, then the
hole-argument does nothing to suggest where to look for a surrogate,
let alone guarantee that there is one. Further, I agree with Earman
and Butterfield (1989, pp. 14–15) in thinking, that abstract algebras
have ontological problems of their own; the surrogate algebra needs
further development. As I read them, the idea is still programmatic.
We can all agree that the speculation is interesting in itself. But in as
much as it is *needed*, that is because the manifold functions as more
than a representation in the orthodox formulation of general relativity: if it goes, it has to be replaced by something else. The recognition

2 This might be obscured by an incautious reading of the diagram on p. 188 of
Earman (1989), a diagram taken from Einstein's correspondence on covariance and
diffeomorphisms. The diagram also appears in Norton (1987).

of a need here (if, indeed, I am fairly construing Earman's discussion this way) concedes that the hole argument *fails* to show that manifolds play only a representing role.

What is the upshot for realism? Suppose that Butterfield and Maudlin are wrong; then realism has to embrace an *S*-indeterminism. It is, nevertheless, an indeterminism of a kind which realism can demonstrate that it is irrational to care about. Its source is the very opposite of mysterious: it springs from the nature of a posit chosen from among others for the very best of *bona fide* reasons of physics. Realism sees manifold properties as real but not as making spacetime a substance; it is the further properties of spacetime, the metric especially, which do that. Further, our grasp of manifold properties goes hand in hand with our understanding of the privileged properties of spacetime. The history of geometry makes clear that the concept of smooth extension and of size and shape are continuous with each other and interdependent. Lastly, realism does not have to face a problem of making sense of spacetime as a representation.

The manifold's role in spacetime theories now stands out boldly, I hope: it provides just those smooth spatial-relational differences which permit smoothly differing field quantities; these then make the aboriginal spatial relations of the manifold determinate in familiar ways (e.g. metrically determinate). It provides the basic and indispensable spatial relations by positing an entity, the manifold space. We still have no other well-established way of providing them.[3]

3 I am grateful for comments made on earlier drafts of this paper from G. S. Hall, Adrian Heathcote, Margaret Rawlinson and Jack Smart.

Part Three

Time and causation

The essays in this part are both about time just by itself and about cause and the relation between them. Time has long puzzled us. Leibniz is the progenitor of one form of an idea that time is an illusion, though founded objectively on what one might roughly call causal relations among monads. This idea has been ably and decisively criticised in Sklar (1974), chapter IV. I take that critique largely for granted here in rejecting the causal theory in general. I take it, as Sklar does, that the idea of cause is dependent on that of time.

Newton said, in the Scholium in *Principia*, that 'absolute true and mathematical time . . . flows equably from past to future without regard to anything external'. It is the idea of flow which Newton evidently felt picked out the essential feature which makes time different from space. Plainly enough, flow is a metaphor for this deep-seated feature, whatever exactly it is. We need to ask what there is about time which gives the metaphor such dominance over us. Clearly it does refer to some deep structural feature for it is difficult indeed to think about time without the metaphors of flow or passage pervading our thoughts, however confused and confusing they are.

The metaphor of flow is tied somehow to the idea of the present as a reference point. But since there is no grip for a metaphorical flow of space, we take ourselves to refer to the present time in something other than the way we refer to the present place. The structure of the difference seems to be this: a flow metaphor will be apt for some dimension *t* just if *t* contains some distinguished reference point *P* such that different *t* items are adjacent to *P* in this dimension at different times. Thus, let *t* be the logical dimension of sequences of thoughts and *P* the focus of one's conscious attention, and one has the metaphor of a flow of ideas; let *t* be a sum of items of production in a factory and *P* the back end of the production line-up, and one has the metaphor of the flow of production. This makes good sense if *t* is not itself time. The idea of a flow of time does puzzle for it has this structure: different items in a dimension of time are adjacent to the reference point of the present at different times. Unless we construe the two occurrences of 'time' in this account of the metaphor as referring to different dimensions of time, the metaphor of time has no purchase since it is trivial. But how does it make sense on that construction? Surely it was not the explicit thought that we could refer to two dimensions of time which drew us to the metaphor. But the most obvious analysis of the metaphor requires a temporal dualism. It needs two orders of time – arguably, an infinite hierarchy of orders. Thus the metaphor sheds no light on the deep structure which it aims to capture and illustrate.

It is easy to see, and it has been much noted, that there is a connection of some kind between the flow of time and the pervasive feature of European languages that they are tensed. No sentence of English is well-formed unless it makes at least some formal link between the content it reports and the time of its utterance. Not all languages have this feature. Its importance has been exaggerated, in my view. Much more significant is the fact that we conceive of the founder members of the world as continuants; that is, as things which are extended in space, have spatial parts and are 3-dimensional, but which have no temporal parts and are not extended in time. It is that ontology which gives point to tense structures in language. Continuants have histories, but events in the histories are related to them not as part to whole but as predicate to subject.

The link between a metaphysics of continuants and our ideas of time is significant. It is continuants which change, and do so precisely because they have no temporal parts but only spatial ones. A thing which has qualitatively different parts is variegated, but this does not entail that it is changing. If however, something has, *as a whole*, a property *P* at one time but lacks the property at another, then this does not mean that it is variegated but that it changes. So the link between time, change and continuants is clear. If there are no continuants then there may be variegation, including variegation in time, but there is no change.

Continuants form the basis of the ontology of common thought. It is hard not to think that they do so because of some ontological bias in common thought which makes it conceive the ontic structure of time as fundamentally different from the ontic structure of space. Places other than the present place do not differ in ontic status from the present, but times other than the present time have a different, inferior ontic status to the present. To an astute philosophical eye these supposed ontic differences make no kind of sense. Yet, in some obscure form, the conviction of ontic differences among times runs very deep indeed. It is almost impossible not to assent to it by accepting the flow metaphor, confused though the assent may be. So the ties between the prominence of continuants in ontology, the conviction that somehow only the present is real, and the flow of time bind us in a circle of ideas which are extremely puzzling for metaphysics.

In particular, historical statements are true, and not just true in history books in the way that it is true in *Hamlet* that Hamlet is melancholy. To find a truth maker for statements about the past one needs an extended time which is enough like space ontologically to sustain them. I think that we find here a confused basis for the metaphor of the flow of time. We want to run two inconsistent pictures of time together, one in which only the present is real, another in which times other than it are real. Running both yields the metaphor as analysed before. There is an ontologically space-like, history-verifying dimension of time, different items of which are *at* the present at different times in some other dimension of time. The present moves across the first ontically space-like dimension during this other order of time and picks out, as ontically distinguished in some obscure way, the times which are at it. No wonder that time proves so confusing and stressful to think about.

We should expect the sources of this confusion to be obvious and pervasive in our experience: obvious because the pull of the metaphor is felt by everyone, pervasive because there is no experience whatever which does not seem to have a 'presentness', a division into past, present and future. So we have a problem of locating the obvious and pervasive in such a way as to explain why, nevertheless, it proves so hard to articulate what these sources are. It is this problem that the following essays mainly try to deal with. There is a quite different problem: what physical facts about the world are the basis and source of the facts which are, in experience, pervasive, obvious and yet, in their peculiar way, so elusive? Presumably, the basis is not differences of ontic status among times since that idea itself is so resistant to clear understanding. The problem of a physical basis is not extensively considered in these essays. I think it is fair to say that it is still not well understood, though very interesting work has been done on it recently, for example in Horwich (1987) and Zeh (1992).

This is the background, sketched rather broadly, against which the two following essays are to be seen. The second of them looks at the idea that the direction of time might be based on the direction of cause by probing the radical possibility that causal relations might hold from later events to earlier ones. There are several approaches to this question, but I think that little light is cast on the problem unless we ask it within a context in which we can see just what the causal link might possibly be. Interesting though Dummett's (1964) reflections on backward cause certainly are, there is a gratuitous puzzle in his example. He imagines a high correlation between the dancing by an elder of a primitive tribe and the bravery of the absent young warriors, the correlation depending on the continued dancing of the elders after the battles are over and the warriors are making their return journey. But this is puzzling independently of the retroaction of the dancing. How could the dancing conceivably cause the bravery even if the temporal order were quite the usual one. Similarly, Mellor's (1981, pp. 177–87) objection to backward cause gives us no indication of what the causal link might be. In his example the putative backward cause has only this attenuated structure: there is a probable correlation between the lighting of certain fires and the later setting of lighted matches to them. Here again, the generality of Mellor's purpose forbids his specifying any detailed, feasible causal link. A statistical correlation is evidence for a causal connection, no doubt. But it never constitutes the link. How is the cause mediated in this example? Mellor proposes a kind of bilking experiment in which one sets matches to half the fires and sets none to another half. But this fails to take account of the full richness which would surround any real causal situation. Consider a familiar example. Pressing a switch turns a light on only in the context of a complex of conditions, intact wiring and a working source of electricity being only the most obvious of them. Now think of turning off the light as a reversed cause of its having been on. One might easily procure a set of examples in which a pressing of a switch turned a light off only half the time because of faulty globes, failure of power and so on. But equally easily, one can find a set in which the light fails to turn on and for the same reasons. This has no tendency to show that there was no causal link in the cases where the light is on. If we could see, in Mellor's example, how the cause was supposed to act, how clear would it be that we could arrange for the later 'causes' to be linked half the time to earlier burnings and half to no burnings? And, if we could, how clear is it that this negates the claim to a causal connection? In the examples I pose, breaking the 'causal' connection is not at all a simple matter. Unless the counter-examples spell the causal conditions out fully they are not sufficient to pose a serious objection. Thus I set my discussion in the context of this question: can we find *actual* instances of the causal relation (analysed in as much detail as possible and assuming omniscience about the conditions which actually obtain) holding between one event and an event earlier than it?

This might seem to conflict with a main conclusion of the first essay, that time travel is impossible as it is usually conceived; that is, so that the time traveller appears in her own past and then proceeds to do things which 'have not happened' – shoot her own grandmother, for instance. But this is not at all the same kind of circumstance, for here the local causal relations are all supposed to be orthodox 'forwards causal' relations. What is problematic here are the non-local relations round the loop, whether the loop be a closed spacetime one, or continuous in the traveller's 'personal time'. I consider just the continuous spacetime loops in this essay.

10 Can time be finite?

1 Introduction

Time *can* be finite, surely. Indeed, we might expect that a finite time would be easier to understand than an infinite one, infinity being looked on, popularly, as a concept before which the mind can only recoil in defeat at any attempt to grasp it. But, of course, that is not how things are. The thought that time, if not the world, is somehow infinite has been the dominant (if not the only) idea about cosmic time in our culture. This excepts the dim, remote past and also the present, in specialized circles, at least. We can begin to explain this state of affairs by pointing to the success of Euclidean geometry as a theory of space. Doubtless, it is easier to think of time as infinite if we first grow used to thinking of space as infinite. But we must find more and deeper presuppositions than this if we are to understand fully the sources of conceptual stress which the idea of finite time is apt to produce in us. I hope to identify at least some of these. For much the most part, I will focus on rather primitive sources of confusion which obstruct the thinking of the generalist rather than the specialist, though some examples push the inquiry necessarily closer toward the fringes of cosmology.

My strategy will be to compare and contrast the problems of understanding finite time with those of understanding finite space. I begin with a broad intuitive distinction. Let us divide finite spaces and times, rather roughly, into two sorts which I will call, simply, open and closed. Open spaces or times, of some number of dimensions (1, 2, . . . n, . . .) are topologically like line segments, bounded areas, bounded volumes or hypervolumes. They come to an end or have boundaries which they may or may not contain. (Thus my use of 'open' and 'closed' is not quite standard, though it is clear enough for my purpose.) Closed finite spaces or times, of some number of dimensions (1, 2, . . . n, . . .), like the circle, the spherical surface and n-dimensional spherical hypersurfaces do not come to an end but close in upon themselves. Though this division between open and closed is exclusive, it is by no means exhaustive, dealing, as it does, with topologically rather simple examples. It is, admittedly, a loose characterisation, but it is asked to bear no real weight in the argument and it is intuitively clear, I believe.

I will argue that differences in the temporal and spatial problems spring on the whole from pre-analytic and rather deep seated convictions about how time differs from space. I will argue that if we are asked to envisage a finite space it seems less objectionable and easier to imagine a closed space than an open one, even if it may not be exactly easy. By contrast, there seem to be fewer difficulties in understanding finite times if they are open than there are if they are closed. I will try to explain why this should be so in the hope of shedding some light on the ways in which pre-analytically we think about space and time.

Kant considered the problem of open finite time and space in the arguments which make up the antithesis of the first antinonomy of pure reason. The firm consensus of modern writers on the arguments is that they are invalid (see, for example, Strawson 1966, pp. 176–85). I will not try to add to this discussion since I do not think that the Kantian arguments focus on the most immediate puzzles, the simpler, but not least intense areas of conceptual stress. Kant's arguments about finite time are quite symmetrical, really, with his arguments about finite space and, while this may reflect a more sophisticated viewpoint than I shall discuss, there are elements which it omits.

2 The beginning of time

Consider open times and spaces. In each case we find a problem about the boundary. But there is a difference. The question which seems most urgent about the temporal boundary when it is a beginning of time is structurally quite different from the urgent question that arises about space. Asked to envisage that time began a finite while ago we may find that we cannot help asking what happened before this, nor help feeling, insistently, that something must have happened before it. Something is structurally spurious about this question, since the question itself (not just an answer to it) refers us beyond the limits envisaged by the hypothesis. The question is raised about a time, which, it (pre)-supposes, precedes the first moment of cosmic time.

The natural question about finite open space is not like this, though the standardly envisaged *answer* to the question may conflict with the hypothesis that space is limited. What we want to ask about an open space is: what happens if we try to throw something at the boundary? The ancient answer to the question was a disjunctive dilemma reducing the hypothesis to absurdity. Either then your spear would pass the boundary, in which case there is somewhere for the spear to go beyond the alleged limits; or it would strike a barrier which could only be a physical and thus material object occupying a space beyond the alleged limits. The problem whether the question takes for granted a space outside the alleged limits is the problem of how to meet this dilemma. On the face of it, the question presupposes only the boundary, some place inside the boundary and the chance of hurling an object in the interesting direction from that place. I take the first horn of the dilemma to be sharp and hard: if the spear passes the boundary, then there *is* space beyond it. We must crumple the second horn. At present, we have no theory which would tell us what happens when a projectile runs out of places to go nor do our theories, as they stand, suggest an expectation. The mind is wont to boggle.

It might be helped, however, by flights of imagination not altogether wild. John Earman has suggested (1977, p. 120) that a boundary to space might be fitted into the broad structure of our present theories of things by acting as a reflecting surface. Presumably the relevant variables change sign when the particles, wavefronts, etc. reach the edge. The question is whether this would attribute causal, mechanical properties to space so that we are really conceiving of it as some kind of brick wall which needs to lie in a containing space in order to have the required mechanical and electromagnetic properties in some intelligible way. However, there is, I think, no need to attribute these physical effects to a material object cause. I have argued elsewhere (§7) that the structure of space might explain quite dramatic and readily observed mechanical effects without its being at all proper to suppose that space thereby acquires mechanical properties or that the component which it adds to the explanation is other than purely geometrical. This shows that we can answer the spatial question without reference to a space outside our envisaged open finite space and without treating space as something other than just a geometrical object. I do not mean to suggest that if space were to play a more properly physical role as spacetime does in General Relativity, it would thereby become like a material object. Explanation by means of spacetime structure in General Relativity is still geometrical explanation. I am arguing that the present considerations are simpler and more straightforward than those in General Relativity (see §7.5). Therefore the possibility of such an answer shows that the question itself does not presuppose a space beyond the finite open space and is not structurally spurious.

The questions that spring so readily to our lips about how time might have begun are not like the spatial question. We cannot answer the temporal questions, say anything which the questioner wants to hear, without falsifying the hypothesis which gives rise to the problem. The solution would seem to be plain. Reject the question. However, to the questioner it seems necessary, inescapable. So, once again, the question is used to reduce the hypothesis to absurdity. Why does the question seem inescapable? It is because two very deep-seated, highly metaphysical, beliefs generally held as necessary truths, though inarticulately, give rise to it.

The first of these metaphysical beliefs is well known, in philosophy at least, as the Myth of Passage of Time (see Williams 1951; Smart 1966; 1967). The image of time's passage or flow is the image of an ontically preferred time (the present) moving along a space-like dimension in which events lie ordered (which I will call cosmic time) *during* a *second dimension* of time (which I will call supertime). The myth of passage is thus tied to deep but generally inarticulate beliefs which give rise to a picture in which there is a dualism (or a still higher ascending regress) of times.(see Broad 1938, vol. II, Part 1, cl. 35, for example). Among these deeper beliefs lie conflicting impulses to treat only the present time as real, yet also to treat statements about the past as true. The existence of the myth and that it is a myth is widely agreed. Its sources remain to be more carefully studied, in my opinion. Prior (1968, cl. 1), for example, sees the myth of passage as springing from rather superficial, rather gratuitous, verbal confusions. Various writers (most notably Smart 1966) attribute it to a mistaken tendency to assimilate events to things and to regard events as changing. While I share (and am indebted to) the latter view I am convinced that the ultimate sources of the myth lie in fundamental facts of temporal perception as contrasted with spatial perception, facts which are still not well understood. Some work has been done on the problem (Broad 1923, chapter 10; this volume §11.6, for example), but one aim of this paper is to add weight to the view that the problem lies deep, that it pervades our thinking about time in a way difficult to escape, and that it is all the more seductive for remaining, generally, inarticulate.

The role of the myth of passage in creating spurious difficulties for our grasp of a finite open time is as follows. We see cosmic time as a finite space-like segment – the dimension in which cosmic events are to lie is a line which is bounded. Across this, during supertime, the present is supposed to move. In supertime, there will be a moment when the ontic nimbus of the present touches the first of the series of events which lies in the space-like dimension, cosmic time. (For easy exposition, I am supposing a first moment of time, i.e. that time contains a lower bounding moment. This is not necessary for the argument, however.) Thus it seems effortless, indeed irresistible, to suppose that there is a moment of supertime which precedes the event of the becoming present of the first moment of cosmic time. Since I am assuming that we do not articulate the structure of the myth (to articulate it is to see, at once, that it poses difficulties) we naturally set no bounds to supertime so that the question what happens, in supertime, to make cosmic time begin is not only possible, but necessary. We might, of course, take supertime as finite, too, but this is a sophisticated thought, hardly likely to occur to us without making explicit the beliefs whose strength, while inarticulate, I am hoping to emphasize.

So what is structurally spurious about the question what occurred before the universe began is not that we are embedding the open finite time of our hypothesis in a wider time which contains it *as a part*, but that we are gratuitously presupposing a whole distinct temporal dimension within which cosmic times are seen as all copresent.

The limits of cosmic time therefore, do not suggest limits to super- (or metaphysical or theological) time. Hence the question seems inescapable, especially when its sources lie among our deeper, unscanned, confused beliefs.

Another pre-analytic belief, which might seem to make it necessary to ask the structurally spurious question about time, is the conviction that every state of affairs has a preceding cause. This conviction is strictly inconsistent with there being a first state of affairs, though not strictly inconsistent with there being a finite time for the history for the universe. To say that there is no first member of a series of finite intervals (earlier members of which are smaller than later members) is certainly not to say that the sum of all the terms of the series which precede a given member is infinite. Like the first pre-analytic, metaphysical belief, the second belief may also be held as necessarily, inescapably true, perhaps (again) the more so the less clearly we articulate it. Hence it, too, might seem to make the spurious question a mandatory one. But, like the first belief again, it does not strictly do so even if we accept the belief. It is easy enough, however, to think of the temporal interval by which the needed cause must precede its effects as never falling short of some given interval, and easier still simply not to notice what one's metaphysical belief strictly is. So the sense that we must ask what preceded a finite cosmic past is understandably vivid. This impulse springs from the myth of passage and its role is, structurally, like the role which the myth plays in our asking what happens before cosmic time begins. The passage of the present through cosmic time reaches the end of cosmic time at some moment of supertime. What, we ask, happens then? But this question *is* a question which refers us to a 'when' in supertime, not cosmic time. It is a structurally spurious question quite analogous to the one about how cosmic time begins. Once again, the impulse to ask the question, which is surely strong, points toward not some peripheral linguistic muddle into which we may fall if we are careless with our usage, but deep-seated tensions among our ideas of time which are founded in the structure of temporal experience. (See Ellis 1955 for discussion of other related issues.)

Parallel problems arise with the conception of a finite future, as one might expect. I shall not repeat what I have said in connection with a beginning for time in the finite past but take it as obvious enough how the arguments would apply.

3 Closed times

Let us turn now to the problem of finite times which are closed on analogy with the sense in which a circle or a spherical surface is a closed space. But, first, what do we find if we look at closed *spaces*?

Spaces with topology Sn (the analogue in n-dimensions of the spherical surface of the sphere in two dimensions) seem to raise a problem of understanding. The affine and metric structures of any of these spaces require us to describe the space as curved. Not surprisingly, this suggests that our closed n-space is curved up within some wider embedding space of higher dimensions ($m + n$ dimensions) in just the way that we think of the spherical surface S_2 as curved up in a higher-dimensional ($2 + 1$ dimensions) Euclidean space. The suggestion is mistaken. The idea of a curvature intrinsic to a space is quite distinct from the idea of its exterior curvature relative to some embedding space. (This distinction is made out in an intuitive way in Nerlich 1994, §§3.1–5.) A space contained in a higher space retains its intrinsic curvature no matter how we bend it about in the higher space, provided that we do not stretch or tear it. It is mathematically convenient, very often, to treat a curved space (or regions of it) as embedded in a Euclidean space of higher dimension because Euclidean geometry is very simple in many respects. But this emphatically does not mean that we can argue that a closed finite physical space must have some wider physical space to contain it and so is not a self-contained, self-subsistent entity.

This seems to be the only problem likely to beset our grasp of closed finite space. Many more problems arise for understanding closed times, and none of them is like the spatial problem just looked at. Before reflecting on these problems let us look at a visualizable model of a universe in which time is closed. Let the toroid in fig. 10.1 represent the spacetime of such a universe with a and b representing time-like curves, a being closed and b almost closed. a is a circle of the toroidal surface and, in fact, a geodesic of it. Different points on a can represent the same place at different times. Lines p and q are segments of spacelike curves which (if produced) would be closed circles and geodesics of the toroid. Different points on p represent different places at the same time and, since p is part of a closed curve, it represents a closed finite space for this universe. Incidentally, the diagram thus understood represents a universe which expands and contracts,

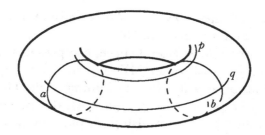

Fig 10.1.

since q, the circumference of the universe at a different time from that shown by p, is larger than p. Crude though such a representation certainly is it gives some support to intuition in envisaging the examples shortly to be discussed. (Such a toroidal spacetime is also discussed briefly in Earman 1977, §3.)

Our simple sketch can hardly be claimed to represent a physically realistic picture of what the universe might be like. Nevertheless, there are live issues in General and even in Special Relativity which present us with essentially the same problems. There is a solution of the General Theory equations due to Kurt Gödel in which certain time-like lines are closed. Actually, spacetime itself is topologically Euclidean in the Gödel model (Stein 1970; Malament 1984), so that the only closed time-like lines are curves, i.e. the trajectories of *continuously accelerated* particles. The conceptual problems posed are not really changed by this fact, however. The discussion of this model and kindred fanciful examples has given rise to a moderately extensive philosophical literature. I will report, briefly, some of these results and make some observations of my own. Similar difficulties are made in Special Relativity by the possibility of tachyons, i.e. faster than light particles. It is, or seems to be, the case that the discovery of tachyons could lead us to the construction of something very like closed time-like loops in spacetime (see Salmon 1975, pp. 122–3). So the issues here do not by any means lie in the world of pure fantasy even though we are not yet confronted with them in practice.

Gödel gives a non-technical account of his argument (1959; see the very accessible account of the Gödel universe in Malament 1984). He describes his case more graphically than I have done, as follows: '. . . by making a round trip on a rocket ship in a sufficiently wide curve it is possible in these worlds to travel into any region of the past, present and future and back again, exactly as it is possible in other worlds to travel to distant parts of space' (p. 560).

In our sketch, the almost closed curve, *b*, represents something like the travels of such an astronaut. His life stretches right round the toroid and comes arbitrarily close to joining itself. The overlapping parts of this curve represent the period when this person might communicate with himself. I do not wish to dispute the possibility of such spacetime structures for the universe.

4 Apparent problems in closed time

Gödel goes on, however, to deduce a paradox. 'This state of affairs seems to imply an absurdity. For it enables one e.g. to travel into the near past of those places where he has himself lived. There he would find a person who would be himself at some earlier period of his life. Now he could do something to this person which, by his memory, he knows has not happened to him' (Gödel 1959, pp. 560–1). He might, in particular, greet his earlier self in, say 1970, with much emotion. Thus in 1970 *x* is both greeted and not greeted as a long-lost self. (I will use the name '*x*-young' to refer to this person in 1970 as he is young and '*x*-old' to refer to him in 1970 after his return as he is old.)

Why did Gödel, and why do we, find it so effortlessly, disquietingly easy to conclude that it is plainly possible for *x*-old to do to *x*-young what did not happen to *x*-young in 1970? Gödel's words are: ' . . . he could *do* something to this person which he knows . . . *has* not happen*ed* to him' (ibid.). My italics stress that it seems quite clear to Gödel that he may use different tenses to refer to the very same time in 1970. I did just the same in the first sentence of the paragraph. Why does neither use jar upon the ear? The tense sequence strongly suggests that we find ourselves ready to see what *x*-old does as occurring later than what is done (or not done) to *x*-young, despite our hypothesis that what *x*-old does by his action is the very same event as what happens (or fails to happen) to *x*-young. What might make that seem natural?

The fly in the ointment is clearly the myth of passage. We add to a perfectly coherent picture of an almost closed time-like curve the idea of a unique, ontologically distinguished time, the present, which can be represented by a curve like *p*. The events which make up curve *b*, which is *x*'s life, are like a circular array of lights, each being lit only when it is touched by the passing nimbus of reality, the present. As the present passes (in supertime) round the circular array of events of cosmic time it comes *back* to an event a second *time* (supertime) and the light is lit *again*. Thus we foist on ourselves the view that *x* is in 1970 *at two* (second-order) *times*: this explains the ready use of two tenses to refer to one time and it explains why we suppose that different things might be true in 1970 first time round from the things that are true second time round. Without the ontological distinctions required to float the idea of the moving present and the whole apparatus of temporal dualism, *nothing* makes it plausible that, in 1970, *x* can be both greeted and not greeted. Without these dubious assumptions nothing lends even an air of intelligibility to any claim that *x* is at 1970 at two times. The ease with which we follow Gödel's footsteps shows how ingratiating those assumptions are, provided only that we keep them unformulated.

We can describe the example consistently, like this. A person *x* is born in 1950, at the dawn of the rocket age. In his twenty-first year, 1970, to the whole world's astonishment, a rocket is sighted approaching the earth from an immense distance and lands safely. From it there alights a 60-year-old astronaut who gives his name as *x*. X tells many a strange tale of the future – or the past. It gets hard to tell which. He seeks out and greets young *x* with enthusiasm telling him that he is his former self. So we are treated to the improving spectacle of *x* beside himself with joy. Old *x* and young *x* see a great deal of each other. Young *x*, who has always harboured astral ambitions, is persuaded by his mentor to train as a spaceman. Later old *x* dies. The combined work of Gödel on General Relativity plus the revolutionary discovery by *Y* of hyper-drive leads to *x*'s departure in a rocket ship on a space-time loop back to 1970. (In my usage, hyper-drive does not produce super-luminal velocities, which I take to be physically absurd and anyhow unnecessary for time travel over long distances. 'Hyper-drive' refers just to the overcoming of engineering difficulties, which Gödel takes a gloomy view of (1959, p. 561), but which, he supposes, allow us to escape the contradictions which otherwise ensue.) After many vicissitudes (but also before them, of course) he arrives, aged 60, to the whole world's astonishment. And so on. The *description* now becomes repetitive, but nothing implies that the events themselves do.

Nothing in this description entitles us to say that *x* is at 1970 at two times. We can say that x is in two *places* in 1970. I see no objection in that, however. In sum, on the question of consistency in Gödel's world, we can say this: it is possible that x-young and x-old are side by side in 1970 and that x-old greets x-young, joyfully, as a long-lost self. It is also possible (though the possibility hasn't been discussed) that x should never meet himself. But it is *not* possible, and nothing in the case suggests it should be, both that x-young is not greeted by anyone in 1970 and also that he *is* thus greeted in 1970 by x-old.

I want to take up, now, the problem of the freedom of action of x-old.

Perhaps x-old would now come to regard himself no longer as a free agent, for it seems that he may recall not merely what he has done, but, now, what he *will* do. As x-young he saw x-old do these things and, as x-old he remembers. A memory of E may both precede and succeed E. But, let us ask not the *subjective* question about x's *view of himself*, but rather these *objective* questions: would his actions really be unproductive of those events for which they suffice, and would his actions be fatalistically bound?

The answer to both questions is: not unless our ordinary actions are unproductive and bound. *All* that could differentiate x-old from x-young in respect of their powers and freedoms, if *anything* could differentiate them, is x-old's epistemic state. Raze the table of x-old's memory and they are precisely in the same predicament or vantage point. For x-old, his head full of remembrances of things future, a characteristic epistemic feature of ignorance of what is about to happen will be lacking. But that is *all* that will be lacking. It is not that, for x-old, the immediately ensuing results of action are *ontically* determined in a way that they are not for x-young. It is not that, for x-old, the future is ontically fixed just as the past is for all of us, whereas for x-young the future is ontically open. There are no such ontic differences for either of them. A future ontically open, in contrast to an ontically closed past, is not needed to make sense of purposive actions. At least, not in any logically worrying sense.

The consistency conditions touched on here are extensively discussed in Lewis (1976), Dwyer (1978), Thom (1975), Clarke (1977), Maudlin (1990b). Clarke suggests that continuity conditions will always ensure that the paradox will be avoided by some non-arbitrary condition. This is closely and illuminatingly examined by Maudlin, who, I think, does not take the point made in the next section based on a remark of Einstein. All these writers take it that ageing can occur right round an almost closed worldline. But it is neither the closure nor the near-closure of the worldline that creates the problem.

5 A problem about time travel in closed time

Despite these arguments for the consistency of time travel stories, the impression is apt to remain that something is wrong with them. I think that this impression is correct. I will argue that, if we were to tell a consistent story about *x*, in full detail, it would be necessary to claim experiences for him which depart drastically from the norms of our temporal experience – so drastically that many would regard the story as conceptually absurd. I do not mean to suggest, at all, that there is, therefore, an absurdity in the idea of a spacetime with Gödelian structure (or any other structure which permits closed time-like lines in spacetime). The oddities or adsurdities lie in what is needed to make entertaining science fiction out of the physical possibility of such structures. What lends *x*'s story the air of something verging on the physically real is that his worldline is continuous and everywhere time-like, so that he travels in space by means which seem not to depart too sharply from current means of travel in space beyond the earth. (Dwyer 1978 appears to take the approach to realism of the story with some seriousness.) However, we should not suppose, even if we live in a world of the appropriate structure for spacetime and could build a rocket ship to go fast enough, that *x*'s journey would be feasible. Gödel's reason for supposing that *x*'s journey is not possible is that it poses insurmountable engineering problems (1959, p. 561). But that can't be the right *kind* of reason to get us out of consistency problems!

Einstein's comment on closed time-like lines in the Gödel universe pinpoints an essential difficulty. Take a point *p* in spacetime. At *p* there is a light cone which divides all spacetime curves through *p* into those which are time-like and those which are space-like (at *p*). Take nearby points *A* and *B* which lie on some everywhere time-like curve through *p*, such that *p* is between *A* and *B* on this curve. Is there some asymmetrical relation which holds between *A* and *B* which would permit us to say that the event *A* precedes *B*? The light cone itself does not provide the relation. Nothing in the spacetime structure round the point, nor in the theory of electromagnetism, gives any basis for claiming that A is before B in some asymmetrical sense of 'before'. However, 'If it is possible to send (to telegraph) a signal (also passing by in the close proximity of p) from B to A but not from A to B, then the one-sided (asymmetrical) character of time is secured' (Einstein 1959, p. 681). That is, we have a basis for claiming that B is earlier than A. 'The sending of a signal is, in the sense of thermodynamics, an irreversible process' (ibid.), a feature which is no doubt connected with the growth of entropy. Einstein asks whether this local solution to the problem of time direction could be extended to the global scale. He concludes that it cannot be if there exists a series of points constituting a time-like line which is closed on itself.

The fundamental structure of Einstein's argument is by no means peculiar to time, let alone to thermodynamics. Consider a closed curve in space. It cannot be that, for every pair of nearby points on this curve, A and B, A is redder than B if and only if A is clockwise from B. No matter what the quality or quantity, it cannot everywhere grow from point to point right round the curve taking the points always in the same geometric sense. Similarly, neither entropy, ageing nor any other factor can everywhere increase from point to point in the same geometric sense right round the time-like curve which we are envisaging. This is a very general form of argument common to spatial and temporal structures, and so easily understood.

Of course in the time-travel story we are interested in, the world-line of x is not closed, though it is almost closed. This fact makes the story interesting because it allows x to communicate with and, more generally, to act causally on himself. If x-old passes a signal to x-young (talks to him, for example) then there is a closed, time-like causal curve in the example, even if no one person, thing or particle has such a curve as its worldline. One rather simple way to show how this casts some kind of shadow across the intelligibility of the story is to suppose that there is a thing (let it be a clock) which has a *closed* history (all the particles which compose it have closed time-like curves as their worldlines) and which x always has with him. To change our tale in a simple way, suppose that x-young enters the time machine and goes on his travels as soon as x-old leaves it, and that the clock is handed from x-old to x-young as they pass at the entry. So x always has a clock with him. We can suppose this 'clock' to be a very crude, simple mechanism. For example, it might be a stick in which x cuts a notch at every mealtime. Now we can certainly say of *x*, consistently, that he ages, accrues memories, scars and so forth. Whatever he gets, by way of traces, he can consistently keep. This is because his world-line is not quite closed. But somewhere, somehow, the stick must lose notches if somewhere, somehow it gains them. For its worldline is closed. We can easily see how it could get notched. But how can it get *un*notched? Not by familiar processes like saving the cut pieces and glueing them back. There must be no trace of the notch, not merely no visible trace, if we are not to run foul of the general sort of inconsistency mentioned before. (David Lewis is well aware of this problem, as I found in discussing the issue with him – a discussion which clarified my arguments in this paper. I do not know whether he would agree with the way I try to extend the argument.) What *x*-old brings back from his journey must *be* the stick which *x*-young departs with. The 'shadow' cast across the intelligibility of the story is suggested by this question: what sense can we make of the fact that, though the watch *x* wears *cannot* always register increasing time, the man who is wearing it can and does.

So far, these remarks about the clock are particular examples of something very general, its generality being well caught in Einstein's discussion of signalling. The relevant facts about watches and sticks are, I hope, immediately graphic. The facts about signalling cut deeper into the problem, however. The asymmetrical direction of time does not require that a signal *is* sent from *A* to *B*. It is given by the mere *possibility* of sending from *A* to *B* together with the *impossibility* of sending from *B* to *A*. If the basis of the impossibility is given in thermodynamics, then the impossibility is not nomological. It is the impossibility of the astronomically improbable. Even if that improbability is what the 'impossibility' comes to, sending a signal from spacetime point *A* to spacetime point *B* as well as sending from *B* to *A* is not a possibility we know how to reckon with. Further, I claim that, in all our experience, this directedness of processes is *regional.* It is more than strictly local, that is to say, even if it may not be global. Thus, if it is possible to send a message from *A* to *B* but not vice versa then given any spacetime points, *C* and *D*, *C* near *A* and *D* near *B*, it is possible to signal from *C* to *D*, but not vice versa. In fact, we take it that relations among time-like connectible events have a common temporal direction over a very large region of spacetime.

I suggest, further, that (within the region) the directedness is quite general over a wide range of processes. That is, in all our experience, if a message can be sent from *A* to *B* but not vice versa, then along *all* time-like lines (in the region) connecting points in the region of *A* with points in the region of *B* the whole class of irreversible processes are similarly directed from the *A* region toward the *B*. Ageing processes in terms of all sorts of phenomena are co-directed.

This takes Einstein's argument up in a quite non-technical sense and extends it in two ways: it claims that the principle applies regionally, not just locally, and that it indicates the directedness, not of particular actual processes, but of a whole range of processes which we may sum up as constituting ageing generally. It is this generalised property which, in our experience, always increases along the same direction of time-like curves. As for this broad sense of ageing, we need not concern ourselves with the question whether the directedness of processes in time rests on a thermodynamical basis, on the fact that our epoch is one in which the universe is expanding or on some other basis. We can rest the argument, I believe, on a quite intuitive understanding of it. Let us apply this to x's journey.

Let us take a finite number of points round the closed time-like Gödelian curve of x's watch. Suppose these points are numbered consecutively from 0 to n-1 and are placed round it at equal intervals (of the proper time of the curve). It is not quite a simple matter what 'consecutively' means when we speak of the ordering of points on a closed curve. Intuitively, it means something like 'in a clockwise sense'. If we wished to be strict, we ought to introduce the four-place relation 'x,y are pairwise separated by v,w'. (A clear account of this may be found in van Fraassen 1970, pp. 69–70.) Let us assign 0 to the point in the history of the watch where x-old leaves and x-young enters the rocket. It seems easy to suppose that x can signal ahead from 0 to 1, from 1 to 2 (when he reaches 2) and so on from j to $j + 1$ (modulo n for positive integers j) for all points thus numbered. That is particularly easy since we suppose that x grows older, accrues and retains information, bears scars and so on as he reaches point j, and moves on to $j + 1$. What Einstein is telling us is that this cannot happen. The spacetime regions within which the Gödelian trajectory lies cannot be so constituted that, at every point j, a message can be sent from j to $j + 1$ (modulo n) but not vice versa. On the contrary, at some stages it must be possible to send a message from $j + 1$ to j and *not* vice versa. That is to say, the direction of time, of ageing, cannot everywhere follow the direction of increasing *j*. Why not? Because this requires a quantity (age) to increase round any of the (presumably infinitely many) closed time-like curves which lie in the region of spacetime nearby the almost closed time-like curve of *x*. This, we saw, is contradictory. Hence, somewhere in *x*'s journey, time runs backwards for *x*, though he himself ages. He watch runs backwards, his stick gets renotched, his rocket motor absorbs fuel from the environment which pushes it, in reverse, through space in some miraculous way, and so on. Because ageing is regional and covers a wide range of phenomena, his experiences include happenings like seeing today the sending of a signal which arrived yesterday (according to *x*'s personal time). This is consistently describable, as I have said, and even imaginable, to some extent, along the lines of seeing a film run backwards. But even if we had reason to believe that the universe is Gödelian (toroid, whatever) and that we could get over the problems of rocket engineering, we could have no reason at all to think that the world might permit *x* to age round his almost closed time-like curve as required to make his tale entertaining in the way we have been considering. This part of his story takes him into realms of experiences the like of which we shall never encounter and which must strike us as strange almost to the point of incomprehension, and which can certainly lay no claim to physical realism.

In standard discussions of *x*-old's encounter with *x*-young (Lewis, Dwyer, Thom) *x*-old must fail (for example) to kill *x*-young, but there need be no particular reason for his failure. Something distracts him, he loses interest, the trigger is somehow not pulled. Which accident fends the contradiction off matters not at all. The strength of the Einstein approach is that accidents will not save the story from collapse into the difficulties focussed upon. It is not that *x* *does* signal ahead from *j* to *j* + 1, nor even that he tries to. The fact that the history of the watch yields a *closed* curve means that no quantity can increase from *j* to *j* + 1 (modulo *n*) at *every* point *j* on the curve. That the curve is time-like and in an arcane geometrical manifold, spacetime, makes no difference to this very simple argument which applies as much to spatial as to spacetime loops. Somewhere on the closed curve ageing must decrease from some *j* to *j* + 1; the timelike direction of signalling, and all that accompanies it, must reverse. That we may *consistently* think of *x* as escaping this reversal does nothing to alleviate the strangeness or to save the realism of his story.

Somewhere, for x, the direction of flow of events generally is opposite to the direction of the flow of his consciousness. For him, time somewhere runs backwards.

Let me return, finally and briefly, to the main theme of the paper. We find that the problems which arise most readily for an understanding of closed time are unlike those involved in grasping the idea of a closed space. The problems of §10.4 were at bottom spurious ones bred by confusions and misunderstandings which beset our thinking about time whatever it be like – open, closed, finite or infinite. The example of closed time simply intensifies problems which are there in any event, whether we notice them or not. The problem discussed in this section is, in its way, spurious too. It arises because the continuity of a closed time-like line suggests that we can understand the idea of a 'journey' round it – of x 'sending himself' from 0 to point 1, point 1 to point 2, . . . j to $j + 1$ . . . and back to 0 in terms of a physics which might be quite realistic (as suggested in Dwyer 1978). No doubt this seems the easier because we do not generally think of increasing time, the widely diversified phenomena of ageing, as an increasing quantity. Once we see that it is, then the problem how to regard closed time-like lines is essentially like simple problems that occur with closed spatial loops, and it becomes, thereby, much easier to understand. Our sense of time as an irrecoverable mystery somewhat abates.

11 How to make things have happened

1 Introduction

Might something I do now make something have happened earlier? This paper is about an argument which concludes that I might. Some arguments (Dummett 1964; Scriven 1957) about 'backward causation' conclude that the world could have been the kind of place in which actions make things have happened earlier. The present argument says that it is that kind of place: that we actually are continually doing things that really make earlier things have happened. The argument is not new (see e.g. Taylor 1964). It sees temporal direction as logically independent of any direction which necessary and sufficient conditions may have, and it sees causal direction as properly deriving from the latter. Thus the directions of time and of making things happen need not coincide and, as it turns out, do not actually coincide in fact. There are examples of events which are sufficient, in a suitably rich sense, for the occurrence of earlier events; hence they make the earlier events happen. My purpose in this paper is to lend support to the argument by filling in some details which 'backward sufficient conditions' may lay claim to. But in one direction I reduce the content often claimed for 'sufficient condition' by omitting all use of modal expressions. This avoids many of the criticisms which have been levelled at Taylor (1964), for example.

The strategy of the argument is simple enough: decompose the causal relation into component relations, replace the component 'later than' by 'earlier than' throughout and find instances of the new complex of relations. In fact, there seem to be many instances. The argument, so interpreted, purports to prove this conclusion: if actions someone performs now make things happen later, then actions someone performs now make things have happened earlier. But the argument presupposes (correctly, I think) a metaphysics of time as a sum of regions among which the present is on a par, ontically, with all the others. The use of 'regions of time' and 'regional theory of time' to refer to what is often called the A-theory or block theory of time derives from Prior (1970). The theory referred to sees different times, like different places, as distinct regions in time, no region having ontic preference over another by virtue of being the present time or place. Time, like space, is the sum of these regions. I prefer this usage as more descriptive and more accurate than others. Though the usage is Prior's, the theory is not. Prior explicitly rejects it in favour of a theory in which only the present is real. There can be no proper relations across time according to the best non-regional or process metaphysics of time, in the pages, mainly, of A. N. Prior. If time is not a sum of temporal regions then there cannot be an ontological parity among the terms of any temporal relation other than simultaneity. Hence cause is not properly a relation and the relational strategy of the argument fails. I think that Prior's account of the temporal syntax of common speech is probably correct, and that he is also correct in supposing that the main features of this syntax spring from pervasive and profound facts about time or about our experience of it. I will argue that the facts in question are not ontological but epistemological, so that the regional metaphysics of time is correct. Nevertheless the epistemological facts may show something about the matter in hand. I will suggest that they require that no one could *intend* to make something have happened. That is quite consistent with the conclusion drawn, obviously.

One serious reservation about the style of approach in this paper is that it strictly presupposes a complete and correct theory of the causal relation. I believe that we still lack this theory, so that I am unable to prove that the fly in the ointment is not some feature of cause. I think, nevertheless, that there is good reason to train the present battery of arguments on the target of time. Partly this is because the argument is consistent with all the theories of the causal *relation* that I know. But, more importantly, our best efforts to produce a relational theory of cause are likely to go on failing so long as we do not resolve our problems about the ontology of time.

The most puzzling thing about cause, I believe, is our pre-analytic view of it as creative. The causing event creates the effected event. That is, we feel (in general) that when the causal event is real the effected event is not real and when the effected event is real the causal event has ceased to be real. This is a dark saying, I admit, but it arises directly from those deep impulses which motivate the non-regional view of time. Something akin to this rather inarticulate belief is connected, I think, with the prevailing but unsolved problem of causes necessitating their effects. What leaves us dissatisfied with Humean theories of cause may not be that the relations offered are too thin *qua relations* to match our idea of cause. It may be that we are looking for something which will capture how the cause in *its* present-ness creates a transition to the effect in *its* presentness. Perhaps it is a pre-analytic idea of cause as the *engine* of time that we miss. If it is, it seems unlikely that analysis can match so inchoate a notion. I think, therefore, that we will not solve the problem of cause till we solve problems of time like the one raised in this paper. I should be well content just to bring this perspective on the problem nearer the centre of our collective attention.

I conclude this introductory section by presenting the bare bones of the argument. Later sections flesh this out.

2 The argument

Variables range over events:

1 $(x) (y) (x$ makes y happen $<\rightarrow x$ causes $y)$ Assumption
2 $(x) (y) (x$ causes $y <\rightarrow x$ is an empirically sufficient condition for $y)$ Assumption
3 $(x) (y) (x$ is earlier than $y <\rightarrow y$ is later than $x)$ Assumption
4 $(x) (y) (x$ is an empirically sufficient condition for $y <\rightarrow y$ is an empirically necessary condition for x$)$ Assumption
5 $(\exists x) (\exists y) (x$ is an empirically necessary condition for y & x is earlier than $y)$ Assumption
6 $(\exists x) (\exists y) (x$ is an empirically sufficient condition for $y <\rightarrow x$ is later than $y)$ 3, 4, 5
7 $(\exists x) (\exists y) (x$ makes y happen & x causes y & x is later than $y)$ 1, 2, 6

The argument is valid, but its premises are not plausible as they stand. This skeletal argument serves a heuristic purpose, however, since it suggests ways in which we can look for examples of making things have happened as we develop richer theories of cause. Assumption 2 is intended to present only the thinnest version of a Humean theory of cause and calls for immediate amplification. The idea of empirically sufficient condition assumed in it is the following. A thing's being F is an empirically sufficient condition for its being G iff (x) $(Fx \rightarrow Gx)$ is true but not logically true. A richer and more careful version may be found in Sanford (1976), pp. 220–2, and problems of extending the idea to distinct events and objects is discussed in Nerlich (1971). As for the remaining assumptions, 1 is obvious, 4 is illustrated and discussed later and 5 may be made plausible by examples – loading a rifle is a preceding empirically necessary condition for the action of firing it.

Is it analytically false that a cause may succeed its effect or that we may make something have happened? Notoriously, there are no clearcut answers to questions about analyticity, and the present question is not among the most straightforward. First, the kind of discussion in the literature about whether causes can be simultaneous with their effects suggests some uncertainty as to which temporal conditions, if any, can fairly be claimed as contained in the meaning of 'cause'. Second, it is not plausible to regard the phrase 'backward cause' as simply a breach of linguistic rule, else we could regard it with the nonchalance with which we accept the phrases 'bachelor girl' or 'grass widow'. If any phrases are strictly analytically true of nothing, these latter phrases probably are. Yet they create nothing like the sense of conceptual outrage that 'backward cause' is apt to produce. Lastly, it may actually be better to think of our deep-seated resistance to the idea of backward causes as like the resistance to an allegedly synthetic necessary falsehood. We seem to feel conceptual outrage at the idea that any event can stand to earlier events in ways at all like the ways in which causes stand to their effects and which lead us to think of causes creating effects. I suggest that this sense of outrage does not spring from an affront to intuition which some relation among events perpetrates. Rather it springs from those (seeming) facts about time that tempt us to say that it flows, that the past is somehow there (though less real than the present) and that the future is nothing at all.

3 Fatalism about the past

The argument that we make things have happened is not an argument that we can change the past. This concession does not trivialise the argument however. We must equally concede that the fact of our making things happen later does not mean that we change the future (cf. Dummett 1964). If you will eat toast for breakfast tomorrow then nothing you or I do now makes it happen that you won't. However, we all assume that something done sometime, now, earlier or later, makes it happen that you *will*. Just so, if you ate toast for breakfast this morning, nothing you or I do now makes it have happened that you didn't eat it. But this is perfectly consistent with the claim that something done sometime, now, earlier or later, makes it have happened that you *did*. When I make some one thing happen, or prevent something else, I never change *what will happen* only *what is going to happen*. Just so, I never change *what has happened*, but always something else.

That use of the 'going to' construction is mildly deviant. To illustrate it further, nevertheless, consider a case where I set an alarm clock to go off at 7 a.m. I shall say that the clock is now going to ring at 7 a.m., though it was not going to do so before I set it. Suppose that, some time later, I press the knob on top of the clock to prevent it from ringing then. I shall say that it was going to ring at 7 a.m. until I pressed the knob on top, after which it was not going to ring. A strictly analogous use of the 'going to' construction is perfectly correct to express *intention* toward the future – the *set* of a human system to perform in a certain way. I extend the use beyond human or animal agents to closed systems generally to express how they are set to perform subsequently. Of course, the 'going to' construction regularly applies to non-human systems, though, standardly, it is then used to express the speaker's certainty about an outcome (see Thompson and Martinet 1969). But dissatisfaction with the grammarian's array of future tenses has been voiced by many philosophical writers (Dummett 1964; Hartshorne 1965; Kotarbinski 1968; Strang 1960 to mention only some modern instances). The literature already contains valuable suggestions as to the logic which such a construction might incorporate (Lehrer and Taylor 1965; Woolhouse 1973). The present idea has three advantages: it is consistent with determinism but does not require it, it modestly extends an existing usage and it is based on the epistemically accessible idea of a local closed system. I shall elaborate it a little further.

Suppose, once more, that I wind and set an alarm clock. Afterwards it is going to ring in eight hours' time, though it was not going to do so before. In the standard cases we say this when we view the clock as a system virtually closed save for human intervention. What counts as a system and as closed here is vague, and the criteria for it anthropocentric. In winding and setting the clock I produce a state of this homely system which is a sufficient condition for the alarm to ring later so long as the context stays closed. To say that the clock was not going to ring before I set and wound it is not at all to deny that prior states of some wider system were sufficient for its later ringing. To say that something is going to happen is to comment on localised causal conditions seen as making up a closed system. It is these systems that we directly change. Similarly, if I later prevent the clock from ringing, I intervene in the system otherwise seen as effectively closed, to destroy the condition, sufficient within that limited context, for the alarm to go off. I press the little knob on top. Thus, to say what is going to happen is not to say what will happen.

There is no expression like 'going to happen' which applies to our actions in their role of making things *have* happened. One is obliged to conduct the argument in a language whose pervasive bent is against the idea that things can be made to have happened. Reasons for this emerge more clearly later. But we can certainly find all the materials for an expression like is 'going to happen' only past-directed. We can give, fairly precisely, what it means to say that something is going to happen, like this:

At t_1 E is going to happen at t_2 *iff* E is a change in some local system, S, and the state just of S at t_1 is sufficient for E at t_2, provided S remains closed between t_1 and t_2.

Nothing in this definiens entails that t_1 precedes t_2. Let us turn to an example. I fire a rifle. Beforehand, a state of the rifle as a closed system was a sufficient condition, just within that limited context, for the rifle to have been loaded earlier. Afterwards, the state of that closed system is not sufficient for the rifle's having been loaded. This is not to deny that later states of some wider system are sufficient for its having been loaded earlier. That the change in state made the loading have happened is a comment on the prior sufficient conditions localised in the system and their later divergence, just as before only reflected in time.

Now for 'preventing something from having happened'. At 11.30 a.m. my watch reads 11.30 a.m. At 2.00 p.m. I change the hands so that the watch reads 1.30 p.m. Seeing the watch as a system closed save for human intervention, some yet later state of the watch, at 3.00 p.m., say, is sufficient, in the closed context, for it to have read 11.00 a.m. at 11.30 a.m., unless it has been altered. Of course I did alter it. I fiddled with the knob that changes the hands. This ought to mean that I prevented it from having been slow at 11.30 a.m.

4 Some theories of cause

I shall look quickly at some prominent theories of the causal relation and try to illustrate very briefly why I think that none of them disables the argument, which still delivers its conclusion. Some writers on cause have faced the problem of 'backward causation' as a challenge to their theories, but they tend to do so in one of two ways. They may appeal to a non-regional metaphysics for time. They may depend on a preference (sometimes implicit as in Geach 1972, or explicit as in Gale 1968, p. 112) for preceding sufficient conditions over succeeding ones and, while this is a congenial preference, it does beg a question against the thesis that one can make something have happened.

The most powerful modern version of the Hume–Mill theory of causality is J. L. Mackie's (1974). Since 2, in the skeletal form of the argument above, is the simplest version of this theory, it seems appropriate to begin with Mackie. That 2 is plainly inadequate is a further reason for looking first for a more satisfactory version of the regularity theory. We need not follow Mackie into the detail of his account, lucid though it is. It will be enough to present some examples which conform to the criteria he lays down. (Mackie introduces the idea that at any time all earlier events, as well as some later ones, are *fixed*. I do not see how this avoids being an amalgam of the causal relation with ontic differences among times, hence an appeal beyond the pure causal relation: 1974, chapter 7).

Consider switching on a light. Here, the closed system is the electric circuit, including the power source. We say that the event of the switch's changing position, or our action in changing it, makes the bulb light up. But this event alone or the state of the switch succeeding it is not sufficient for its lighting. The state of the circuit sufficient for the bulb to be lit is complex and the position of this switch after the change is only part of the complex. Moreover, other bits of circuitry might have been manipulated to make the bulb light up. It might be a circuit with several switches. Here, causes are plural and complex. What we single out as making the bulb light up is neither sufficient nor necessary, but merely part of a complex state, one among a number of possible alternatives, sufficient for the light to go on.

Consider, now, switching the light off, but regard it from a new perspective, as switching the light so as to *have been* on. We get a perfect reflection of what we just saw. The closed system is the circuit, including the power source. The state of the switch preceding the action is not alone sufficient for the bulb's having been lit. The state of the circuit which is sufficient is quite complex, and the position of this switch before the event is only part of the complex. Moreover, other bits of circuitry might have been manipulated to make the bulb have been lit. It might be a circuit with several switches and we might have changed another of these. The situation here is the time-reflection of the action ordinarily called switching the light on.

Hume thought that continuity and contiguity are part of our concept of cause. Whether or not he was right in this, we can easily find instances of making things have happened that embody both principles. I think it is important to show that the air of magic which surrounds examples of 'backward causation' like those of Dummett (1964) and Scriven (1957) might be dispelled.

My last case aims at a more detailed time-reversed analogy with setting an alarm to go off at a definite time. Some telephones have a mechanism which makes a buzzing noise when the phone is left off the hook and disengaged. This gets louder as the time goes on. The action in question is the one usually called replacing the receiver and correcting the oversight. But see it instead, as making the phone have been disengaged twenty-five minutes ago. Or, even better, we may say that it makes the phone have been correctly on the hook twenty-five minutes ago and earlier. The point of the example is to catch something which is there in the ordinary alarm clock case. That is, I fix things so that something, the alarm ringing, is going to happen at a definite time later. I initiate a continuously developing process in the closed system which ends with the event in question. In short, the point is continuity, contiguity and continuous development. Similarly in the phone case, I hang up with the noise at a quite definite level. The state of the closed system of the phone immediately before my action is sufficient for it to have been disengaged just twenty-five minutes ago. I terminate a process (let's call it a continuously *en*-veloping one to remind us of our new perspective) which has been going on in the closed system of the phone and which begins with the disengagement.

The rifle and the switch example already meet quite strict conditions of continuity and contiguity. But the rifle case is rather like setting a booby trap. You can be sure what will happen when it goes off, but not when it will go off. We know the rifle was loaded sometime, but not which time. The switching is like an *immediate* consequence: it is an immediate presequence. So the telephone example fills in some useful details which we find in the relations involved in simple instances of a cause.

Kneale and others (Beauchamp 1974, sections 8, 10, 17) have argued that causal laws have some kind of necessary status. A rather detailed attempt to provide a basis for the necessity has been offered by Popper (ibid., sections 12, 13) though he has not claimed that the attempt succeeds. The suggested necessity is best understood as a complication added to the pure regularity theory as Mackie, for example, expounds it. This is especially clear in Popper's explication. Consequently, I find nothing in the suggestion, nor in the attempt to explain it, which is inconsistent with our idea of how to make something have happened.

In *singularist* theories, cause is explained without appeal to natural regularities. C. J. Ducasse (Beauchamp 1974, section 21) argues for the simplest version of a singularist theory, claiming that cause directly relates concrete events, independently of whatever regularities there may be. In order to discover the cause of a change we need find only the single preceding event which is sufficient for, or necessitates, the change. Ducasse does not explain why only *preceding* events should make things happen. In fact, his theory of cause fits our examples of making things have happened in much the same way as the regularity theory fits them. This is exactly what we should expect, granted Davidson's argument that Ducasse's theory is compatible with the Humean theory (ibid., section 26).

Richard Taylor (1966) and H. R. Harré and E. H. Madden (1975) have produced more complex singularist theories. They give a central role to *agency* and to the *powers* or *capacities* of particulars (including persons). Harré and Madden (pp. 114–15) make the concepts of agency and power central to defining the direction of cause. They correctly point out that temporal priority is not logically connected with causal priority, the latter idea getting sense from the idea of a 'powerful particular'. I find no reason in their account of power, why the rifle, the electric circuit and the telephone in the earlier examples cannot be regarded as each having the power or capacity to make the relevant event have happened. However, Taylor (1966, pp. 31–5), appeals to the non-regional metaphysics of time to define a direction for cause, thus going beyond cause as a relation.

Lastly, there are theories of cause which make it an anthropocentric or epistemological concept. There is a recipe theory of cause (Beauchamp 1974, Part 4) and a theory that locating a cause is locating a theoretical explanation (ibid., Part 6). In both theories some element of human interest or capacity overlays purely objective relations among things or events; or it may overlay purely logical relations among various sentences, including law-like sentences, which describe the things or events. In no version of either theory is there any new suggestion that something about these objective or logical relations should give us pause in accepting the argument that things can be made to have happened. If the anthropocentric overlap gives cause a direction in time then that is simply an epistemological or subjective fact, quite powerless to prevent us from seeing the things and events themselves as objectively related in the way the argument requires.

David H. Sanford (1976) argues for criteria which impose a direction on conditionship, so that the symmetry of conditionship (as expressed in 4 of the skeletal argument) fails. After much careful argument Sanford produces this definition (p. 205):

P is a *sufficient condition* for Q iff

(1)   P is sufficient for Q and

(2)   there are admissible circumstances in which everything necessary for P is necessary for Q.

Here, clause (1) uses 'sufficient' much as I have been using 'empirically sufficient condition' and (2) requires that in admissible circumstances there should be no back up or standby conditions for Q were P to fail. I have no quarrel at all with this definition, but I fail to see how it refutes e.g. the claim that changing the hands at 2.00 p.m. makes the watch read 11.30 a.m. at 11.30 a.m. in the example above in §11.3. In correspondence Sanford agrees that the definition does not refute the claim, though he has not accepted the diagnosis offered in this paper.

This theme suggests that it is linked to a theory in which cause is constituted by the truth of counterfactual conditional statements (see Lewis 1979). Here again, the theory seems to deliver what our intuitions ask for. For consider again the alleged ways of causing my watch to read 11.30 a.m. at 11.30 a.m. This might come about in the ordinary way by my changing the watch at 11.00 a.m. from reading 10.30 a.m. to reading 11.00 a.m. I set it correctly some time before 11.30 a.m., in short. The following intuitive conditionals seem to make this plainly different from the 'fake correction' after the event when, at 12.00 p.m. I set the watch to read 12.30 p.m. We would all agree that:

If I had not changed my watch at 11.00 a.m. it would not have read 11.30 at 11.30

But we would certainly not agree that, in the time reversed case:

If I had not changed my watch at 12.00 p.m. it would not have read 11.30 at 11.30

Neither conditional appears to show a question-begging preference for preceding factors over succeeding ones. So here there seems to be a definite refutation of the causal claim, even though infected by the unclarity and dispute to which counterfactual condionals are prone.

But these conditionals conceal just that question-begging preference. They assume a temporally slanted view of change; more precisely, a slanted view of what it is for a change *not* to occur. First let us represent the watch changes by the following ordered pairs of events:

11.00 change = <watch reads 10.29 at 10.59; watch reads 11.01 at 11.01>
12.00 change = <watch reads 11.59 at 11.59; watch reads 12.31 at 12.01>

It is the most natural thing in the world to read the conditional antecedent 'If the . . . change had not occurred' as envisaging a (contrary to fact) state of affairs in which the actual ordered pair is replaced by another which differs just in its *second* member. But if one is seriously bent on not assuming the usual preference for time direction, one ought to contemplate readings of the conditional antecedent as envisaging a (contrary to fact) state of affairs in which the actual ordered pair is replaced by another which differs just in its *first* member. Then we would read 'the 12.00 p.m. change did not occur' as

<watch reads 12.29 at 11.59; watch reads 12.31 at 12.01>

rather than the standard reading of the phrase, namely

<watch reads 11.59 at 11.59; watch reads 12.01 at 12.01>

A similarly deviant reading of 'the 11.00 a.m. change did not occur' is plainly open to us. *Of course*, the proposed deviant readings violently affront our intuitions, but it is just our intuitions which are in question. Clearly, the deviant readings yield immediately two new conditionals, given the circumstances as before:

The watch would not have read 11.30 at 11.30 if the 12.00 change had not occurred.

The watch would still have read 11.00 at 11.30 even if the 11.00 change had not occurred.

But if these conditionals are sustainable on non-question-begging readings of the counterfactuals, we have not succeeded in making the appeal to counterfactuals show that the 11.00 a.m. change is, while the 12.00 p.m. change is not, a sufficient condition for the correct reading at 11.30 a.m. (The last two paragraphs are inserted from my (1979), pp. 9–10, where the issue is discussed at greater length.)

Lewis's treatment of the temporal asymmetry of counterfactual dependence is different from this. The preference for forward-looking rather than back-tracking conditionals does not rest, as I have argued, on the way we formulate the antecedents in these particular cases. It rests instead, Lewis argues, on a contingent fact about our world, or perhaps only about the region of the world we inhabit. This underlies the ways in which we settle the ambiguities which surround the counterfactual supposition and the ways in which we find the possible world nearest or most like our own. The fact is that there simply are many many more sufficient conditions for the occurrence of an event after it has happened than there ever are before it. Certainly events can be overdetermined by conditions which precede it, though this is not very common and when it does occur, there never seem to be more than a few such precedent overdeterminants. But there characteristically are multitudes of succeeding overdeterminants for any event. On this fact the asymmetry of counterfactual dependence rests.

Our concern with this clearly relevant and interesting point must be whether or not it forbids us to take later events as related to earlier ones just as causes are, save for the temporal precedence of cause. It does not forbid us. We might build into the definition of cause the condition that a cause must not lie on the same temporal side of the effect as the majority of conditions which are sufficient for it. One can see a point in such a definition. Causes, so defined are salient sufficient conditions in a way in which the quasi-causes that succeed an event might seem swamped by their fellows. Further, it seems to join hands with Sanford's requirement of the admissible absence of back-up conditions.

But I claim that, for our present purpose, a significant feature of this example overrides the multiplicity of later sufficient conditions. For one of the succeeding conditions may be preferred over others. The change in the watch *alters a local closed system* which otherwise remains isolated during the relevant periods, both before and after the change. Plainly the watch does develop (envelop or what you will) from the changed state to the target state in a causal way. The intuitive causal direction corresponds, of course, with the direction in which overdetermining *evidence* lies. But, again, this does not seem to be the kind of asymmetry we want in order to disable *the change in the position of the hands* as irrelevant to the target event.

In all these examples there is a common problem, to be dealt with later, which emerges clearly in the watch case. The manipulation of the watch by an agent precedes the actual change in its hands and thus falls (causally and temporally) between the change at 2.00 p.m. and the correct reading at 11.30 a.m. I try to deal with this worry in §§11.7–8.

5 Cause and time: the syntax of temporal expressions

There are two reasons why it might still look as if cause is the snake in the grass. First, none of the theories sketched in the last section commands the warm assent of almost all of us. The door stands wide open for some quite new direction of attack on the problem and, even more obviously, open to fresh developments of the theories already extant. Clearly, we cannot overlook the possibility that new insights into cause might cast a wholly different light on the problem of making things have happened.

Even so, it is reasonable to lay the trouble at the door of time rather than cause. For one thing, time is notoriously problematic because of our awkward tendency to give different ontic standings to different times. For another, if there are relevant disanalogies between making something happen and making something have happened, they ought to show up in the examples. The ones offered could hardly be more familiar or transparent. While there are disanalogies, I, at least, cannot see how they offer us the allegedly missing element. If the directions of cause and time lie in something pervasive and obvious in experience, part of our problem will go unanswered if the difference springs only from some highly theoretical feature remote from observation.

There is a second reason for thinking cause is the snake in the grass.

The diagnosis which points the finger at cause is not altogether wrong, though this is only because the idea of cause is itself infected with the disease which makes time so problematic. If we were to probe deeper into the maze of a general theory of cause and of making things happen, which avenues would call most urgently for us to explore them? Surely, they are the ideas of cause as a necessary connection and of every event as needing a cause. I do not mean that our theory should incorporate and justify these ideas, but we do need to grasp what prompts such deep intuitive faith in them. Every young philosopher has his or her personal battle with belief in the necessity and universality of cause, and few of us remain content with our cease-fire lines. Once we turn to these problems, however, the strategy of producing examples must fail us. We will not find the necessity and the key to universality in *particular cases*, as Hume saw from the beginning. I suggest that previous attempts to shed light on this dim area may have been misguided in trying to locate the source of necessity in the causal *relation.* But it need not lie there. Quite the contrary. It may well lie in the way *people ordinarily conceive* of cause as a phenomenon in time, and this means that cause is not conceived as a relation at all. Someone might argue as follows.

Causes make and create; they bring what they cause *into being.* That is the heart of causal necessity. Each event must be pulled from the limbo of non-being into the ontic sunshine. The cause must *make* its effect, not merely relate to it. This aspect of cause is missed when we see it as a relation, for a relation holds only between things ontically on a par. But when the cause occurs, the event caused *is not,* yet. When it occurs, the cause is no longer. This binds the concept of cause to an orbit within the metaphysics of time. So cause is not a more primitive concept, useful in probing the foundations of time. The problem of cause *is* the problem of time.

Something tangible lies behind these vague remarks. Prior (1968, chapters 1, 2, 8; 1967, chapter 1) and also Geach (1972, ext. 10.2) have argued that the syntax of temporal expressions is quite different from the syntax of spatial ones. Let us accept this syntactic argument at least provisionally. Spatial expressions are characteristically *relational.* Furthermore, on the one hand, our spatial vocabulary has nothing like tense in it and, on the other, nothing like the part–whole relation in space applies to continuants in time and their histories. These are global features, not just of our language but of nearly any language. I think that it may well be misguided to replace temporal expressions by relational ones in the style of taken reflexive theories of syntax (as examined in Gale 1968, chapter 11). However, I believe that all the philosophical goals of this latter programme of syntax can be reached by a thoroughly plausible programme in semantics.

One brief example of how continuant syntax (see Johnson 1964, vol. III, chapter 7) works might not go amiss, though it is hardly unfamiliar. The words to watch are 'whole', 'part', 'same' and 'is', meant as identity. A busy tourist can see the whole of Stonehenge in a morning. He can't see all of it by seeing all of one monolith; he must see every spatial *part* – every stone. He does not see any Druidical rites performed there, though it *is* a thing, the very *same* thing, in which these rites were performed. Yet he does not miss any part of Stonehenge, nor did the Druids see a part of it which he does not see. He sees the very thing they saw: the whole of Stonehenge.

I sum up a great deal of careful and ingenious argument on the syntax of temporal expressions in these main points: the syntax of temporal discourse is that of sentence operators, not of temporal binary predicates using singular terms which refer to events. Tenses make no *references* to time or events, they are sentence-forming operators on sentences. Some examples: we can deal with all tense structures by tense operators 'Past', 'Present' and 'Future' and interactions of these. They take sentences as operands. There are constructions like 'John's arrival was earlier than Bill's', but the primitive sentence structures are those like 'John arrived *earlier* than Bill did', 'Nero fiddled *while* Rome burned' and 'I watched *as* the plane landed'. In all of these, the temporal expressions are binary sentence operators, not two-place temporal relations taking singular terms. The ontology of all primitive sentences is exclusively a continuant ontology. There is no primitive reference to events. Event expressions and temporal relations are all to be parsed away as linguistic shadows.

If there really are no temporal relations, then there is no causal *relation*. So far as I know, neither Prior nor Geach has drawn this conclusion explicitly. Clearly, it is a momentous one, not only for the enterprise in hand. It might seem that the basic forms of causal sentence are shown in these examples. 'The jolt made the vase fall' or 'Her singing caused the glass to shatter'. Aren't these relational sentences? In each case, the subject is a singular term naming an event. But the object is not a singular term in either sentence, nor could it fill the blanks in the expression ' — preceded . . .', which is a temporal relation, if any is. I suggest that the admittedly very simple verbs 'make' and 'cause', in these contexts at least, are never primitives of syntax. The primitive forms are exemplified perhaps in these: 'The vase fell because it was jolted' or 'Her singing shattered the glass' or 'She shattered the glass by singing', or in sentences like this: 'Jack caused the vase that it falls' for 'Jack caused the vase to fall'.

6 Cause and time: the semantics of temporal expressions

Though I am fairly well persuaded by the Prior–Geach theory of temporal syntax, I cannot accept the theory of semantics that goes implicitly or, in Prior's case, explicitly, with it (Prior 1970). The theory is that tense operators are like the operator 'possibly' and several oth-

others to be mentioned. They have the function of removing the assertive force of the sentence on which they operate. Thus 'There are giant animals' is straight false comment on the world, whereas the operator-bound sentences 'Possibly there are giant animals' or 'Future (there are giant animals)' function, first, to signal that no straight comment is made. Now, the prefixes 'In Greek mythology . . .' or 'In *Gulliver's Travels* . . .' yield truths when they operate on the sentence just mentioned. So does the operator 'Past . . .', and so perhaps does 'Future . . .' or 'In the future . . .'. The first two prefixes can misleadingly suggest that the operand sentence is true in some wider region or universe which somehow contains the world of straight comment. The operator would then be seen as a concealed relational expression. Just so, the temporal prefix can suggest that the present is merely part of a wider reality, a universe of time in which the present is merely one box or region related to other past or future boxes or regions. Tense would then express this relation. But Prior believes this is wrong in both cases. The idea of the present and of the real are one. Each of 'Present', 'Now', 'Really', 'Actually' and 'It is the case that' is an *identity* operation on sentences. The operator's function is pragmatic; it draws attention to the straightness of the comment. It contrasts with comments as to what was or will be so, on the one hand, and with what might be so or what is merely so written or said, on the other. The regional interpretation of operators other than the identity functions is misleading in that 'it minimises, or makes a purely arbitrary matter the vast and stark *difference* that there is between the real and every form of unreality' (Prior 1970, p. 245). A regional view of time, on which the main argument depends, ignores this difference.

I find this theory of the semantics of temporal expressions incredible. On Prior's theory, I see no way of making sense of present *evidence* for past or future tense statements, nor of our prospective and retrospective activities, like planning and recollecting, and others that depend on these. The theory itself 'minimises, or makes a purely arbitrary matter the vast and stark difference' between history and any sort of fiction or conjecture, on the one hand, and between memory and any sort of imagination on the other. Perhaps the sharpest difficulty is to give a clear, intelligible account of how past and future may be thought to differ from each other in ontic status, while each yet retains a different ontic status from the present (good accounts of this difficulty may be found in Fitzgerald 1968; 1969). It is exactly this dim ontic difference between future and past which our argument jams us hard up against. For it is just because the past is somehow *there*, in the ontic twilight, that we can't create it and just because the future is nohow there that we can *create* it. Prior himself saw that the no-regions theory of time conflicts with Special Relativity (Prior 1970, pp.132–4; see also Mellor 1974). His proposed solution – trivialise Special Relativity to mere epistemological significance – asks far too high a price.

However, I repeat that I do tentatively accept Prior's theory of the *syntax* of familiar temporal expressions. It cannot be an accident that spatial expressions which certainly concern a system of regions, enter discourse in ways that differ so radically from those in which temporal expressions enter it.

The *structure* of the semantics of tense and modal logics is simply neutral between a regional metaphysics and Prior's. The structure is simply a set of worlds, ordered by some relation of accessibility (or precedence for tense logic). We cannot regard this array of worlds as an array of regions of a wider universe in the case of the semantics of modal logic – or so I think. But there is no reason why we cannot do so for tense logic. Of course, there remains a critical need for us to explain why temporal syntax should take the elaborate form it does, if a syntax like the simple one for spatial expressions is semantically possible.

Summing up the paper at this point, I have argued for two conclusions. *First*, expressions which are temporal conjunctions, like 'while' and 'earlier than' nevertheless have the *semantics* of temporal relations, so there really are temporal relations, including those of cause and making happen. *Second*, the picture already painted of the relations involved in our idea of making something happen is substantially correct and adequate; since we find instances of the temporal inverses of these relations appropriately conjoined, some events do make earlier ones have happened. I now want to go on to argue, *third*, that the facts on which the tense-and-continuant structure of temporal discourse rests are of epistemological, not ontological, importance. Their epistemological role nevertheless explains why we are inclined to give them ontological weight.

7 An asymmetry in time: traces and memories

After this digression, let us return to regarding time as a sum of regions and the argument in §11.1 as establishing that some events make others have happened. Now, there is a marked asymmetry in time as a sum of regions, one much commented on (e.g. in Gold 1962; Popper 1956; 1957; 1958; Reichenbach 1971). It is the asymmetry of traces. I shall describe traces so as to highlight their influence on us epistemologically rather than so as to prepare them for later inclusion in some theory aimed at explaining them. Roughly, then, traces are enduring states of objects (or systems of objects) which *resemble* an object involved in the event of their imprinting. Traces may, instead, mimic the event of imprinting, or be a means of echoing it. Again, they may simply be remnants that mark the place of some occurrence. Footprints are traces of the first sort, signatures (which trace a pen's trajectory) or sound recordings (films etc.) are of the second sort, while dead embers or merely fire-blackened stones are of the third sort. The existence of a trace is a sufficient condition for the occurrence of the event of interaction. So that while foot and footprint resemble each other symmetrically, only the existence of the latter guarantees the contact of foot and sand at some time. Though traces never precede their imprinting events, we can say what they are without referring to that fact. How does the asymmetry of time in respect of traces bear on the argument so far? Imperfect confidence in the current physical theories of the asymmetry (e.g. in Horwich 1987; Zeh 1992) prevent unqualified answers here. However, the preoccupations found in the present literature on causation may excuse the claim that trace asymmetry has nothing at all to do with the problem how one event makes another happen (or have happened). None of the paradigms of cause looked at so far has any clear relevance at all to the question of traces. A moving billiard ball is not a trace of the collision which moves it. A loaded rifle is not a trace of the loading, nor is a wound alarm clock or a glowing light bulb a trace of the events that bring these states of affairs about. Of course, there are obvious differences between these cases and the contrived examples in sections 2 and 3 of making something have happened. These differences may be vaguely related to traces. For instance, a light filament is always hot just after the globe has gone out, but seldom just before it. But facts like this play no part in theories of cause in the existing literature (as we see it reflected in Beauchamp 1974, for example). Though traces and related phenomena may allow us to describe how states of a system later than a given state characteristically differ from states earlier than the given one, they say nothing illuminating about why actions or events cannot make things have happened. Or rather, as I shall argue later, they do so only indirectly by way of memory and the epistemological theory of intentional action that I will defend. I will try to show that the idea of making something happen is anthropocentric – a projection of our temporal experience onto the world.

As far as I know, only Reichenbach (1971) and those who have developed his theory, (e.g. Grünbaum 1973; Hinckfuss 1975; Smart 1973) relate cause at all closely to traces. But, as I understand this theory, it has no tendency to show that cause is not reversible, if we regard cause as a concept which applies at the level of fundamental physical law. On the contrary, the theory is expressly designed to be consistent with the thesis that the fundamental laws of physics (and thus its fundamental processes) are time-symmetrical. In short, if fundamental micro events make later micro events happen, then they make earlier micro events have happened just as often and in exactly the same ways.

Still, the macro version of this theory does bind causality tightly to the temporal asymmetry of traces, so we must look at the consequence of this for the argument of the paper so far. Reichenbach shows how the spatial ensemble of thermodynamic branch systems can be expected to develop their changing entropy states in the same direction in time (1971, §14; see Grünbaum 1973; Horwich 1987; Zeh 1992) for more developed formulations). It is overwhelmingly improbable that a branch system carrying a trace will develop in a direction opposed to the majority of the systems at the time, thus making the trace borne by the branch system a *pre*-trace. A cause is now to be regarded as a coarsely perceived macro event of interaction which explains the trace, or low-entropy state of the branch system. The fact that there are no backward causes is purely a statistical matter; indeed we must bear in mind the extremely remote possibility that it is not a fact at all. Furthermore, outside our thermodynamical epoch (globally in spacetime) it is not unlikely that there are other epochs where the thermodynamic trends are time-reversed. Of course, this would not be apparent to inhabitants of these epochs since they would be swept along, as we are, in the direction of the prevailing entropy tide. In sum, whether we understand this theory as about cause as a fundamental micro concept or as a merely approximate macro concept, it allows us to speak of reversed-causation, even if in a different style from that of the main bulk of this paper.

This Boltzmann–Reichenbach theory has yet to win broad assent (see Earman 1974; Horwich 1987). My own reservations about it reflect my uncertainty as to just what is meant by a system's 'branching off' and by its 'rejoining' wider systems. More specifically, I think that something more needs to be said about those events of interaction which distinguish a trace proper, like a footprint, from other kinds of branch system states of low entropy, for example Reichenbach's sun-warmed rock embedded in snow (see Smart 1968). But it is unnecessary to pursue these worries here.

Memories are traces. This gives an inescapable temporal direction to every experience we have. Memory gives us a detailed, systematic, intuitive picture of what has happened, especially in the immediate past. Like perception, memory has a large component of unverbalised, unverbalisable information. (There is no mystery about this. Reflect just on the vast superiority of a photograph or a recollection of a face to the best, most complete police description of it.) In the very short term – I mean a few seconds – memory appears to be nearly perfect. Some of this information is encoded (not necessarily verbally) and retained as long-term memory. The information is not just dense, but systematic too. I not only remember a pleasant walk, but that I took it on Sunday, after lunch and before dropping in on friends. (These remarks apply just to redintegrative memory; see Hilgard and Atkinson 1967, section 12.)

Let me stress again that memory is the possession of *non-verbal semblances* of what has happened. A fossil is something much more than a very good description of a long-dead plant. That the plant was *itself* imprinted on the rock gives us a sort of vicarious possession of it, in our possession of the fossil. Similarly, recollection of what has happened gives us a kind of vicarious possession of the past. The vanished plant is somehow *there* in the fossil and the by-gone event somehow *there* in memory. This surely says something notable about our tendency to give the past a reality which we deny to the future.

Yet to explain this tendency is not to justify it. The existence of a trace at t_1 which records an imprinting event at t_0, does not lend the t_0

event a reality at t_1. Nor is the t_0 event any less (or any more) real at t_1 when it has no traces, than at t_0, when it has some. Events neither have nor lack reality at times other than the time when they happen. At least, this is so given a regional view of time. Similarly, it would be absurd to argue that Chairman Hua is more real in the Chinese embassy in Canberra, where they hang his picture, than in the Russian embassy, where they don't. Hua is really in China, and it is plain nonsense to talk about his reality in other places he doesn't inhabit. That I now recall breakfast this morning is no ground at all for thinking it now more real than the dinner which I assume, but do not absolutely know, that I will eat this evening.

We are bent on considering whether our temporal syntax draws its strength from epistemological facts in our experience of time or whether it rests on the ontic standing of the present, the past and so on. The argument from memory is familiar enough to let me have been brief with it. The superior force and vivacity of perception over memory are just as obviously relevant to this epistemological theory. Most of the cash value of 'force' and 'vivacity' can be got by seeing them as metaphors of the superior richness in information that perception has over memory. But what needs further explanation, I think, is why perception is only of the present time, but not only of the present place.

Well, first, it is strictly false that perception is only of the present time. The light from stars comes to us just as surely across the years as it comes across the parsecs. Yet if we watch an occultation of Ganymede by Jupiter, we see it *as* occurring at a distance but not *as* occurring some minutes ago. Perception never presents itself as of things across time but only as of things across space. Again, and this is surely connected closely, we often see things much bigger than ourselves *as* much bigger: indeed we may see mountains, for example, as dwarfing our bodies by their colossal size. But no single act of perception can take in events and processes which last longer than the act itself. (The idea of a single act of perception is vague, but I think not too vague for my point.) More strictly, it is just that a single act of perception cannot take in a process *as* longer than itself. For, in fast motion films, we do see in a quite short time what happened over a longer one. But we do not see it *as* that, but only as rapid motion. Outside film, if an object were to move so fast toward us as to produce a significant Doppler effect toward the violet, we might again see

lengthy processes in a short time. We *can* see large temporal exten-
sions in brief. The small visual angle of distant objects would seem to
correspond, roughly, with the rarely encountered time compression
in perception just mentioned. But this gives us no counterpart of perspective in our perception of events and processes. In spatial perception there is a rigid connection between visual angle, size and distance of what is perceived. In time perception we lack any rigid connection between time-compression, actual duration and distance in
time of what is seen. I suggest that this prevents us from seeing events
either as at a temporal distance or as lasting longer than our perception of them lasts. Clearly the fact that the round-trip time for light
over domestic distances is imperceptible plays a role in prompting
our belief that we do not see across time. I feel quite sure that these
are incomplete suggestions. It is a crucial issue for an epistemological
theory of the idiosyncrasies of time, and any light thrown on it in possible future discussion would be valuable.

Here again, crucial disanalogies between spatial and temporal perception go far to illuminate why we are inclined to say that only the
present is real. Other places are real. I see across them; I see things at
them; I see things that fill arbitrarily large regions of space. But other
times are not real. I do not see them. Either I simply remember them
or I have no direct knowledge of them at all.

8 The epistemology of action

If we consider just intentional action, that is to say actions under a
description according to which they are intended, it seems that we
should not want to say that we can intend to make something have
happened. I will conclude by glancing briefly at this matter, in order
to point out that it does not seriously weaken the conclusion of the
arguments already given. Intentions are formed and carried out by
action in a distinctive epistemological climate. Since the epistemological climate is also a temporal one, we can argue that actions can never
be *intended* to make something have happened. Of course, this does
not entail that our actions don't make things have happened, all the
same. All our actions have countless consequences which we do not
intend, or even notice. The presence of people in a room makes the
temperature of the air rise slightly. A person's speaking imposes a cer-

tain rhythm on the rain of tiny blows from air molecules on the walls of the room. We make these things happen, but usually we are quite unconscious of it.

No doubt there is a law that we cannot formulate intentions, but only attitudes, towards happenings we know of directly. Direct knowledge is to be cashed as perception or memory, each of these being given some time-symmetric explanation, for example, by laying stress on unverbalisable information. No doubt there is a converse law that intentions are directed *by* what we know, especially by what we know directly. Both theses are supported, for example in Armstrong (1968). The most obvious facts are the most likely to be overlooked. Let us remind ourselves of some of these.

To act at all, we need to know what is round us, how we are oriented toward it, what is the disposition of our body and limbs and the sizes and masses of things to be grappled with, which things are at rest, which in motion, how these are currently moving, and so forth. How much we depend for the simplest tasks on knowledge of this kind and how much it directs us is easily gathered by the simple expedient of trying to get along with closed eyes. All this knowledge is gained from perception and the sort of near-perfect very short-term memory that colours our every experience of the world.

Now apply these ideas to the simplest kinds of continuous activity that go to make up our actions, even though we usually regard our actions atomistically. To fire a rifle calls for the simple continuous activities of moving my finger towards the trigger, curling it round and pulling. To switch the light I must guide my hand toward the switch and press in a quite definite direction. Consider the not very primitive, but still rather illuminating activity of maintaining a moving car in the correct position on the road. This activity is almost wholly governed by flow of information about the immediately preceding position and speed of the car on the road, on the position of the steering wheel and of the accelerator pedal. Given a steady speed and a position of the car on the road at a time, the position of the steering wheel determines where the car has immediately been just as surely as it determines where the car *will* immediately be. Nevertheless, the purposive activity is directed exclusively at where the car *will* be. First, in any period of the activity, especially very short ones, the later states of the driver's mind are richer in information about the journey in that period than earlier states are. Second, at any stage of the period, dense unverbalised information is available only about that stage and preceding stages, especially immediately preceding ones, and never about immediately succeeding ones. Every basic activity, every simple deliberate movement, occurs in a context of information which lends it a marked and inescapable orientation in time.

What these reflections suggest is that we cannot claim that we are able to direct our purposive activities toward what has happened, at least, not in any robust sense of 'direct our purposive activities'. My intention to fire the rifle so that it has been loaded or to switch the light so that it has been on must *precede* the action which makes it have happened. The intention lies on the same temporal side of the action, so to speak, with what the action is supposed to be aimed at bringing about. The intention issues in subsequent purposive activity, and it is this, or the event in which it culminates, which actually brings about the preceding event. This yields a conceptual incongruity: the activity by which we make the event have happened does not fall between the event and the formulating of the intention. In fact, the intention is simultaneous with the state of the light's being on in the second case. In the first case, the intention to fire the rifle falls *between* the loading and the firing. That is also true of my watch having been wrong and connecting the phone. This ordering of intention, action and event never occurs when we make things happen later. I conclude that setting out to try to make something have happened is not a clearly intelligible project. But, let me stress again, this leaves the main argument untouched. It never talks about intentions and plans, only about what is made to happen by actions in particular and events in general. That is not only a matter of what is made to happen later. Making things to have happened is a package deal with the regional view of time, and that is my reason for urging you to buy it.

In conclusion, the main upshot of the paper is not best seen, perhaps, as a commentary on our largely unnoticed causal capacities in respect of preceding events. More important is the moral that the concept of cause is powerless to solve the problems posed by the concept of time. The fundamental laws of physics present our most careful, best-established and most sophisticated understanding of time. Notoriously, nothing in these laws endorses the idea of the flow of time, nor of a direction which is basic to our conception of it. Nor are these laws causal (in the sense of singling out causes) even when they are deterministic. The concept of cause is not a fundamental one and cannot illuminate the darker corners in our understanding of the fundamental concept of time.

Bibliography

Alexander, H. G. (1956). *The Leibniz–Clarke Correspondence.* New York, Barnes and Noble.

Armstrong, D. M. (1968). *A Materialist Theory of the Mind.* London, Routledge.

Barnes, A. (1979). 'Cosmology of a Charged Universe', *Astrophysical Journal* 227: 1–12.

Beauchamp, T. L. (1974). *Philosophical Problems of Causation.* New York, Dickenson.

Bennett, J. (1970). 'The Difference Between Left and Right', *American Philosophical Quarterly,* 7: 175–91.

Bilanuik, O. and Sudarshan, E. (1969). 'Particles Beyond the Light Barrier', *Physics Today,* 22: 43–51.

Bridgman, P. W. (1962). *A Sophisticate's Primer of Relativity.* Middletown, Conn. Wesleyan University Press.

Broad, C. D. (1923). *Scientific Thought.* London, Kegan Paul.
(1938). *Examination of McTaggart's Philosophy.* Cambridge, Cambridge University Press.

Butterfield, J. (1987). 'Substantivalism and Determinism', *International Studies in the Philosophy of Science. The Dubrovnik Papers* 2: 10–32.
(1989). 'The Hole Story', *British Journal for the Philosophy of Science* 40: 1–28.

Cajori, F. (ed.) (1947). *Sir Isaac Newton's Mathematical Principles of Natural Philosophy and his System of the World.* Berkeley, University of California Press.

Clarke, C. J. (1977). 'Time in General Relativity', in J. Earman et al. (eds.), *Foundations of Space-Time Theories*. Minneapolis, University of Minnesota Press.

Coleman, R. A. and Korté, H. (1982). 'The Status and Meaning of the Laws of Inertia', *Philosophy of Science Association Proceedings* 6: 257–74. (1993). *Philosophical and Mathematical Foundations of Spacetime*. Dordrecht, Reidel.

Demopoulos, W. (1970). 'On the Relation of Topological to Metrical Structure', in M. Radnor et al. (eds.), *Minnesota Studies in the Philosophy of Science*, vol. IV: Minneapolis, University of Minnesota Press.

Dummett, M. (1964). 'Bringing about the Past', *Philosophical Review* 73: 338–59.

Dwyer, L. (1978). 'Time Travel and Some Alleged Logical Asymmetries Between Past and Future', *Canadian Journal of Philosophy* 8: 15–38.

Earman, J. (1970). 'Who's Afraid of Absolute Space?' *Australasian Journal of Philosophy* 48: 287–319.
(1971). 'Kant, Incongruous Counterparts and the Nature of Space and Space-Time', *Ratio* 13: 1–18.
(1972). 'Notes on the Causal Theory of Time', *Synthese* 24: 74–86.(1974). 'An Attempt to Add a Little Direction to "The Problem of the Direction of Time" ', *Philosophy of Science* 41: 15–47.
(1977). 'Till the End of Time', in J. Earman et al. (eds.), *Foundations of Space-Time Theories*. Minneapolis, University of Minnesota Press.
(1979). 'Was Leibniz a Relationist?' in P. French et al. (eds.), *Midwest Studies in Philosophy*, vol. IV: *Studies in Metaphysics*. Minneapolis: University of Minnesota Press.
(1986). *A Primer on Determinism*. Dordrecht: Reidel.(1989). *World Enough and Space-Time*. Cambridge, Mass., MIT Press.

Earman, J., Glymour, C. and Stachel, J. (eds.) (1977). *Foundations of Space-Time Theories. Minnesota Studies in Philosophy of Science*, vol. VIII. Minneapolis, University of Minnesota Press.

Earman, J. and Norton, J. (1987). 'What Price Substantivalism: The Hole Story', *British Journal for the Philosophy of Science* 38: 515–25.

Ehlers, J., Pirani, F. and Schild, A.
(1972). 'The Geometry of Free Fall and Light Propagation', in L. O'Raifeartaigh (ed.), *General Relativity*. Oxford, Oxford University Press: 63–84.

Einstein, A. (1953). 'Geometry and Experience', in H. Feigl and M.

Brodbeck (eds.), *Readings in the Philosophy of Science.* New York, Appleton-Century-Crofts.

(1959). 'Remarks to the Essays . . .' in P. Schilpp (ed.), *Albert Einstein: Philosopher-Scientist.* New York, Harper Torchbooks: 663–88.

Einstein, A, Lorentz, H. A., Minkowski, H., and Weyl, H. (1923). *The Principle Of Relativity.* New York, Dover.

Ellis, B. D. (1955). 'Has the Universe a Beginning in Time?' *Australasian Journal of Philosophy* 33: 31–7.

(1971). 'On the Conventionality of Simultaneity', *Australasian Journal of Philosophy* 49: 177–203.

Ellis, B. D. and Bowman P. (1967). 'Conventionality in Distant Simultaneity', *Philosophy of Science* 34: 116–36.

Feinberg, G. (1967). 'Possibility of Faster-Than-Light Particles', *Physical Review* 159: 1089–1105.

Fitzgerald, P. (1968). 'Is the Future Partly Unreal?' *Review of Metaphysics* XXI: 421–46. (1969). 'The Truth about Tomorrow's Sea Fight', *Journal of Philosophy* 66: 307–29.

Friedman, M. (1972). 'Grünbaum on the Conventionality of Geometry', *Synthese* 24: 219–35.

(1977). 'Simultaneity in Newtonian Mechanics and Special Relativity', in J. Earman et al. (eds.), *Foundations of Space-Time Theories.* Minneapolis, Minnesota Studies in the Philosophy of Science.

(1983). *Foundations of Space-Time Theories.* Princeton: Princeton University Press.

Gale, R. M. (1968). *The Language of Time.* New York, Humanities Press.

Geach, P. T. (1972). *Logic Matters.* Oxford, Oxford University Press.

Geroch, R. (1971). 'General Relativity in the Large', *General Relativity and Gravitation* 2: 61–74.

(1972). 'Einstein Algebras', *Communications in Mathematical Physics* 26: 271-75

(1978). *General Relativity from A to B.* Chicago, University of Chicago Press.

Geroch, R. and Horowitz, G. (1979). 'Global Structure of Spacetimes', in S. Hawking and W. Israel (eds.), *General Relativity: An Einsteinian Centennial Survey.* Cambridge, Cambridge University Press. 212–93.

Giannoni, C. (1978). 'Relativistic Mechanics and Electrodynamics Without One-Way Velocity Assumptions', *Philosophy of Science* 45: 17–46.

Glymour, C. (1972). 'Physics by Convention', *Philosophy of Science* 39: 322–40.

(1977). 'The Epistemology of Geometry', *Noûs* 11: 227–51.

Gödel, K. (1959). 'A Remark about the Relationship between Relativity Theory and Idealistic Philosophy', in P. Schilpp (ed.), *Albert Einstein: Philosopher-Scientist.* New York, Harper Torchbooks.

Gold, T. (1962). 'The Arrow of Time', *American Journal of Physics* 30: 403–10.

Goldhaber, A. and Nieto, M. (1971). 'Terrestrial and Extraterrestrial Limits on the Photon Mass', *Reviews of Modern Physics* 43: 277–95.

Graves, J. (1971). *Conceptual Foundations of Contemporary Relativity Theory.* Cambridge, Mass., MIT Press.

Grünbaum, A. (1967). 'The Denial of Absolute Space and the Hypothesis of Universal Nocturnal Expansion: A Rejoinder to George Schlesinger', *Australasian Journal of Philosophy* 45: 61–91.

(1968). *Modern Science and Zeno's Paradoxes.* London, Allen and Unwin.

(1973). *Philosophical Problems of Space and Time,* 2nd edn. Dordrecht, Reidel.

Harré, H. R. and E. H. Madden (1975). *Causal Powers.* Oxford, Blackwell.

Hartshorne, C. (1965). 'The Meaning of "Is Going to Be" ', *Mind* 24: 46–58.

Hawking, S. and Ellis, G. F. R. (1973). *The Large Scale Structure of Space-Time.* Cambridge: Cambridge University Press.

Heywood, P. and Redhead, M. (1983). 'Non-Locality and Kochen-Specker Paradox', *Foundations of Physics* 13: 481–99.

Hilgard, E. R. and Atkinson R. C. (1967). *Introduction to Psychology.* New York, Harcourt Brace.

Hinckfuss, I. (1975). *The Existence of Space and Time.* Oxford, Oxford University Press.

Horwich, P. (1987). *Asymmetries in Time.* Cambridge, Mass., MIT Press.

Janis, A. (1969). 'Synchronism by Slow Transport of Clocks in Non-Inertial Frames of Reference', *Philosophy of Science* 36: 74–81.

(1983). 'Simultaneity and Convention', in R. Cohen and L. Laudan (eds.), *Physics, Philosophy and Psychoanalysis: Essays in honor of Adolf Grünbaum.* Dordrecht, Reidel: 101–10.

Johnson, W. E. (1964). *Logic.* New York, Dover.

Jones, R. (1980). 'Is General Relativity Generally Relativistic?'

Philosophy of Science Association Proceedings 2: 363–81.

Kotarbinski, T. (1968). 'The Problem of the Existence of the Future', *Polish Review* 13: 7–22.

Lacey, H. M. (1968). 'The Causal Theory of Time: A Critique of Grünbaum's Version', *Philosophy of Science* 35: 332–54.

Lehrer, K. and Taylor, R. (1965). 'Time, Truth and Modalities', *Mind* 74: 390–8.

Lewis, D. (1976). 'The Paradoxes of Time Travel', *American Philosophical Quarterly* 13: 145–52.

(1979) 'Counterfactual Dependence and Time's Arrow', *Noûs*, 13: 455–76.

Locke, J. [1700] (1975). *An Essay concerning Human Understanding*, ed. P. Nidditch. Oxford, Oxford University Press.

Lucas, J. R. and Hodgson, P. E. (1990). *Spacetime and Electromagnetism.* Oxford, Oxford University Press.

Mackie, J. L. (1974). *The Cement of the Universe.* Oxford, Oxford University Press.

Malament, D. (1977a). 'The Class of Continuous Timelike Curves Determines the Topology of Spacetime', *Journal of Mathematical Physics* 18: 1399–1404.

(1977b). 'Causal Theories of Time and the Conventionality of Simultaneity', *Noûs* 11: 293–300.

(1984). '"Time travel" in the Gödel universe', *Philosophy of Science Association Proceedings* V.2: 91–100.

(1985). 'A Modest Remark about Reichenbach, Rotation, and General Relativity', *Philosophy of Science* 52: 615–20.

Maudlin, T. (1990a). 'Substances and Spacetime: What Aristotle Would Have Said to Einstein', *Studies in the History and Philosophy of Science* 21: 531–60.

(1990b).'Time-travel and topology', *Philosophy of Science Association Proceedings* v.1: 303–15.

Mehlberg, H. (1935). 'Essai sur la Théorie Causale du Temps', *Studia Philosophica* 1: 119–260 and 2: 111–231.

Mellor, D. H. (1974). 'Special Relativity and Present Truth', *Analysis* 47: 74–6.

(1980). 'On Things and Causes in Spacetime', *British Journal for the Philosophy of Science* 31: 182–8.

(1981). *Real Time.* Cambridge, Cambridge University Press.

Misner, C., Thorne, K. and Wheeler, A. (1973). *Gravitation.* San

Francisco, W. H. Freeman.

Mortensen, C. and Nerlich, G. (1978). 'Physical Topology', *Journal of Philosophical Logic* 7: 209–23.

Mundy, B. (1983). 'Relational Theories of Euclidean Space and Minkowski Space-Time', *Philosophy of Science* 50: 205–26.

Nerlich, G. (1971). 'A Problem About Sufficient Conditions', *British Journal for the Philosophy of Science* 22: 161–70.

(1976). 'Quine's "Real Ground" ', *Analysis* 37: 15–19.

(1979). 'Time and the Direction of Conditionship', *Australasian Journal of Philosophy* 57: 3–14.

(1982). 'Pragmatically Necessary Statements', *Noûs* 7: 247–68.

(1994). *The Shape of Space*, 2nd edn. Cambridge, Cambridge University Press.

Newton, I. (1962). *Unpublished Scientific Papers of Isaac Newton*, ed. A. Rupert Hall and Marie Boas Hall. Cambridge: Cambridge University Press.

Norton, J. (1987). 'Einstein, The Hole Argument and the Reality of Space', in J. Forge (ed.), *Measurement, Realism and Objectivity*. Dordrecht: Reidel: 153–88.

Poincaré, H. (1952). *Science and Hypothesis*. New York, Dover.

Popper, K. (1956). Letters to *Nature* 177: 538; 178: 382.

(1957). Letters to *Nature* 179: 1297.

(1958). Letters to *Nature* 181: 402–3.

Prior, A. N. (1967). *Past, Present and Future*. Oxford, Oxford University Press.

(1968). *Papers on Time and Tense*. Oxford, Oxford University Press.

(1970). 'The Notion of the Present', *Studium Generale* 23: 245–8.

Putnam, H. (1969). 'On Properties', in N. Resher et al. (eds.), *Essays in Honour of Carl G. Hempel*. Dordrecht, Reidel: 235–54.

Quine, W. V. O. (1960) *Word and Object*. Cambridge Mass. M. I. T. Press.

(1976). 'Whither Physical Objects?' in R. Cohen et al. (eds.), *Essays in Memory of Imre Lakatos*. Dordrecht, Reidel: 497–504.

Recami, E. and Mignani, R. (1974). 'Classical Theory of Tachyons', *Rivista Nuovo Cimento* 4 (series 2): 209–90.

Reichenbach, H. (1958). *Philosophy of Space and Time*. New York, Dover.

(1971). *The Direction of Time*. Berkeley, University of California Press.

Rindler, W. (1977). *Essential Relativity*, 2nd edn. New York, Springer.

Robb, A. (1914). *A Theory of Time and Space*. Cambridge, Cambridge University Press.

Rosen, N. (1980). 'Bimetric General Relativity and Cosmology', *General Relativity and Gravitation* 12: 493–510.

Salmon, W. (1969). 'The Conventionality of Simultaneity', *Philosophy of Science* 36: 44–63.

(1975). *Space, Time and Motion: A Philosophical Introduction.* Encino, Calif., Dickenson.

(1977a). 'An "At-At" Theory of Causal Influence', *Philosophy of Science* 44: 215–44.

(1977b). 'The Curvature of Physical Space', in J. Earman et al. (eds.), *Foundations of Space-Time Theories.* Minneapolis, University of Minnesota Press.

(1977c). 'The Philosophical Significance of the One-Way Speed of Light', *Noûs* 11: 253–92.

(1982). 'Replies to Critics', in R. McLaughlin (ed.), *What? Where? When? Why?* Dordrecht, Reidel.

Sanford, D. H. (1976). 'The Direction of Causation and the Direction of Conditionship', *Journal of Philosophy* 73: 193–207.

Schlesinger, G. (1967). 'What Does the Denial of Absolute Space Mean?' *Australasian Journal of Philosophy* 45: 44–60.

Schrödinger, E. (1963). *Space-Time Structures.* Cambridge, Cambridge University Press.

Scriven, M. (1957). 'Randomness and the Causal Order', *Analysis* 17: 5–9.

Sklar, L. (1974). *Space, Time, and Spacetime.* Berkeley, University of California Press.

(1977a). 'What Might Be Right about the Causal Theory of Time', *Synthese* 35: 155–71.

(1977b). 'Facts, Conventions and Assumptions in the Theory of Space-Time', in J. Earman et al. (eds.), *Foundations of Space-Time Theories.* Minneapolis, University of Minnesota Press.

(1981). 'Time, Reality and Relativity', in R. Healey (ed.), *Reduction, Time and Reality.* Cambridge, Cambridge University Press.

(1985). *Philosophy and Space-Time Physics.* Berkeley, University of California Press.

Smart, J. J. C. (1966). 'The River of Time', in A. Flew (ed.), *Essays in Conceptual Analysis.* London, Macmillan. 213–27.

(1967). 'Time', in P. Edwards (ed.), *The Encyclopedia of Philosophy.* New York, Macmillan and Free Press.

(1968). *Between Science and Philosophy.* New York, Random House.

(1969). 'Causal Theories of Time', *Monist* 53: 385–95.

Stein, H. (1970). 'On the Paradoxical Time Sructures of Gödel', *Philosophy of Science* 37: 598–601.

(1977). 'Some Philosophical Prehistory of General Relativity', in J. Earman et al. (eds.), *Foundations of Space-Time Theories*. Minneapolis, University of Minnesota Press.

Strang, C. (1960). 'Aristotle and the Sea Battle', *Mind* 69: 447–65.

Strawson, P. F. (1966). *The Bounds of Sense*. London, Methuen.

Synge, J. L. (1956). *Relativity: The Special Theory*. Amsterdam, North Holland.

Taylor, R. (1964). 'Fatalism', *Philosophical Review* 71: 45–66.

(1966). *Action and Purpose*. Englewood Cliffs, NJ, Prentice-Hall.

Thom, P. (1975). 'Time-Travel and Non-Fatal Suicide', *Philosophical Studies* 27: 211–16.

Thompson, A. J. and Martinet, A. V. (1969). *A Practical English Grammar*. Oxford, Oxford University Press.

Van Fraassen (1969). 'Conventionality of the Axiomatic Foundations of the Special Theory of Relativity', *Philosophy of Science* 36: 64–73.

(1970). *An Introduction to the Philosophy of Time and Space*. New York, Random House.

Weinberg, S. (1972). *Gravitation and Cosmology: Principles and Applications of the General Theory of Relativity*. London, Wiley.

Weyl, H. (1952). *Space, Time and Matter*. New York, Dover.

Whewell, W. (1840). *The Philosophy of the Inductive Sciences, Founded upon their History*. London. J. W. Parker.

Williams, D. C. (1951). 'The Myth of Passage', *Journal of Philosophy* 48: 457–72.

Winnie, J. (1970). 'Special Relativity Without One-Way Velocities', *Philosophy of Science* 37: 81–99, 223–38.

(1977). 'The Causal Theory of Spacetime', in J. Earman et al. (eds.), *Foundations of Space-Time Theories*. Minneapolis, University of Minnesota Press.

Woolhouse, R. S. (1973). 'Tensed Modalities', *Journal of Philosophical Logic* 2: 393–415.

Zeeman, E. (1964). 'Causality implies the Lorentz Group', *Journal of Mathematical Physics* 5: 490–3.

(1967). 'The Topology of Minkowski Space', *Topology* 6: 161–70.

Zeh, D. (1992). *The Physical Basis of the Direction of Time*. New York, Springer.

Index